Food habits and consumption in developing countries

Food habits and consumption in developing countries

Manual for field studies

Adel P. den Hartog

Wija A. van Staveren

Inge D. Brouwer

ISBN-10: 90-76998-94-9
ISBN-13: 978-90-76998-94-7

First published, 2006

Wageningen Academic Publishers
The Netherlands, 2006

Preface

This manual addresses food habits and consumption patterns in developing and emerging countries to better understand the nutritional situation of their populations. It comprises practical information on conducting field studies.

Part One of the manual provides insight into the dynamics of food habits and consumption and the accompanying socio-economic and cultural dimensions. Part Two gives practical information on conducting small-scale studies to be carried out within the framework of a nutrition issue, including the collection of data on food habits and the measurement of food intake. This manual on food habits and consumption is intended for professionals with practical or academic training and who are involved in various types of food and nutrition programmes and related activities. It can also be used as a handbook in food and nutrition training courses at professional and academic levels.

The first manual on food habits and consumption was published in 1983 by Pudoc in Wageningen and was updated in 1988 and 1995 by Margraf Verlag in Germany. During the last decade, the food and nutrition situation in the developing world has changed dramatically and new ideas and insights on food and nutrition issues have emerged. Thus, the need arose to write a complete new manual based on new evidence and insights. The material presented is further based on the experiences of the authors while working with the Division of Human Nutrition of the Wageningen University and Research Centre, WUR, Netherlands, bilateral programmes in the field of research and training, and international agencies such as FAO and UNICEF.

Many thanks are due to those who have helped in preparing the manual, in particular Mr Michael Price (editorial work), Ms Janneke van Wijngaarden (technical support), Mr Jan Burema (statistical advice), and the staff of Wageningen International (formerly IAC) of the WUR. The publication of the manual would not have been possible without the generous support of the Dutch foundation *Stichting voor Sociale en Culturele Solidariteit* in Zeist, founded by the late Dr. Ir. F.D. Tollenaar, food scientist (1915-1999).

Adel P. den Hartog

Table of contents

PART THREE: APPENDICES

PART ONE:

FOOD HABITS, FOOD CULTURE, AND CONSUMPTION

1. Introduction

1.1 The nutritional context

Nutrition is recognized as playing a crucial role in attaining development, as reflected in the eight Millennium Development Goals (SCN, 2004). Numerous deficiency diseases continue to persist in the developing world, especially in the rural areas, as a result of essential nutrient deficiencies in the daily diet. These now coexist with an increasing presence of diet-related chronic diseases and adult obesity, which until now was only seen in industrialized and developed countries (Shetty, 2002). Both conditions can co-exist, and this co-existence is often referred to as the double burden of nutrition, particularly in emerging countries.

The trends in child malnutrition vary greatly from one region to another; the situation in sub-Saharan Africa is especially alarming. The number of underweight pre-schoolers is increasing and will continue to do so unless strategic moves to improve the situation are implemented. Pre-school stunting shows the same pattern; the prevalence and numbers of wasted pre-schoolers are projected to increase in various parts of Africa. Pre-school malnutrition trends reflect the deteriorating situation in many sub-Saharan African countries, where the poverty rate has increased, HIV/AIDS is having devastating impacts, conflict persists, and gains in agricultural productivity as the key driver of overall economic growth remain elusive (SCN, 2004). In Asia, the numbers of malnourished children are still very high in absolute terms, and trends are decreasing slowly over time.

In addition to the concern of whether people consume enough energy (and protein), there is an increased concern about the adequacy of micronutrient intake, the vitamins, minerals, and essential fatty acids that help the body function. Although data is scarce, available estimates for micronutrient deficiencies are alarming. More than half of pregnant women and school-age children suffer from iron deficiency anemia, as do more than 40% of non-pregnant women and pre-school children. Some 100-250 million pre-school children alone are affected by severe vitamin A deficiency and 740 million people are affected by goiter, a symptom of iodine deficiency (Ruel, 2001). Evidence is increasing that deficiencies of other micronutrients like zinc, folic acid, calcium, and essential fatty acids may be as serious. Rapid population growth and a dramatic increase in urbanization further challenge food supplies and changing food demands. On the other hand, the prevalence of overweight and obesity is also increasing at alarming rates among both the rich and poor, and among rural and urban populations. This may lead to an increase in cardiovascular diseases, diabetes type II, and other chronic diseases. The accelerated reduction of malnutrition (both under- and overnutrition) is also needed where the trends in household food security and poverty are moving in the wrong direction.

A great number of interrelated and often obscuring factors cause malnutrition, which are multi-faceted in nature. For a better understanding of this, it is useful to take the different levels of causes into account. UNICEF has developed a conceptual framework on the causes of malnutrition. (see Figure 1.1).

The basic causes of malnutrition are the socio-economic structure and the prevailing political system of a country, town, or region. To a large extent, these factors determine

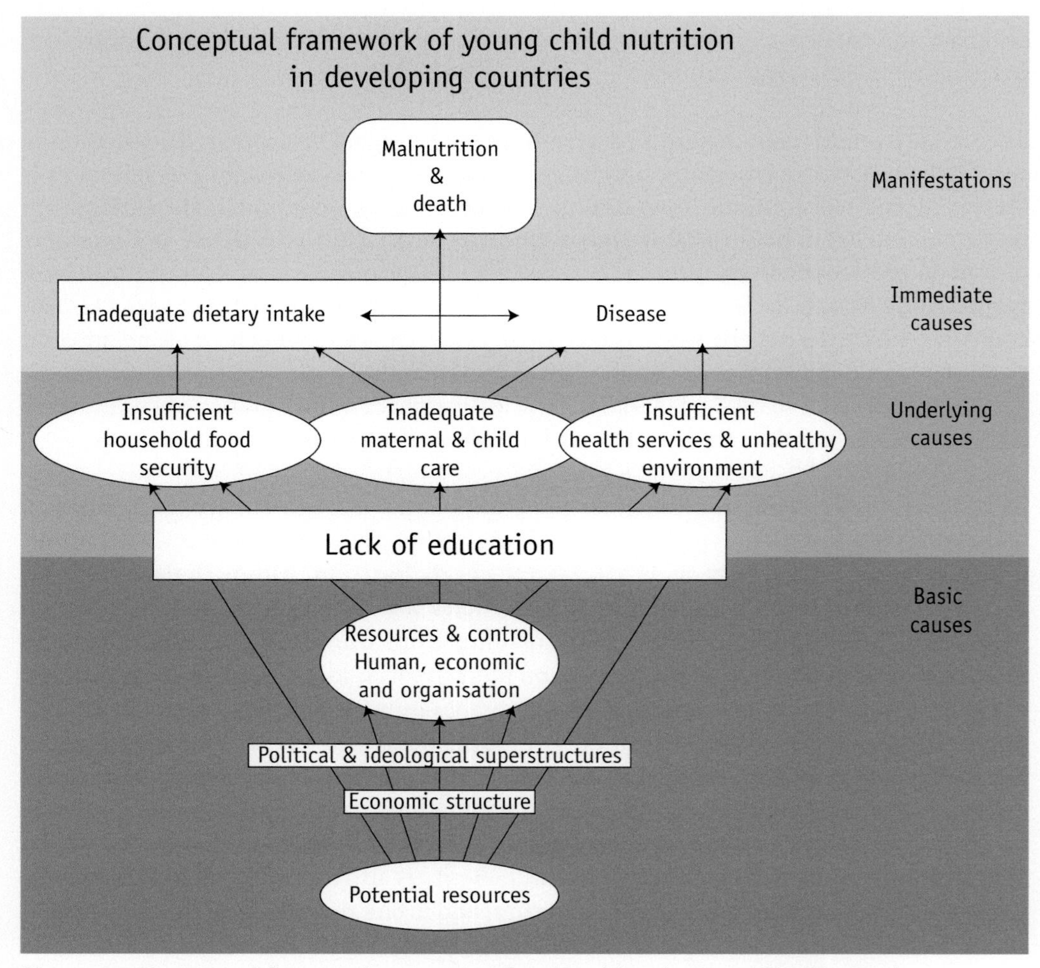

Figure 1.1. Conceptual framework on causes of malnutrition (source: UNICEF, 1990).

how accessible resources such as land and food production, housing, and income are to the household and its different members. The underlying causes of malnutrition are basically household food insecurity, insufficient health services, and an unhealthy environment. Nutritional care or how a household copes with scarce food and health resources are key determinants in maintaining healthy nutrition. There is empirical evidence that some poor households do better than other poor households. It is amazing how some poor households are capable of making the utmost use of their meagre resources. Needless to say, there are limits to what nutritional care can do in deprived areas. The immediate causes of malnutrition are often a combination of inadequate food intake and infectious diseases.

1.2 Aim of the manual

Many of the constraints encountered in nutrition intervention programmes are rooted in a weak understanding of how people deal with their food, which in turn is closely linked with the prevailing socio-economic conditions and food culture. To understand the nature of a nutritional situation, its causes, and possible options for solutions, it is essential to have information on the socio-economic and cultural data influencing food habits, or how people use the food available to them.

The aim of this manual is to provide an outline on how to collect data on food habits and food consumption. It is intended to be a manual not only for the staff of food and nutrition programmes, planners such as nutritionists, community health staff, agronomists, and extension officers, but also for students in training courses at the professional and academic levels. Many professionals may feel the need to collect data on food habits and food consumption during the course of their work, as such data will enable them to get a better insight into the nature of a nutritional problem and finding ways of solving it.

This manual provides practical advice for small-scale studies on how to collect data on food habits and food consumption. Such data may be collected for three inter-related key objectives:

- To gain insight into the socio-economic context of food and nutrition of a community for which little or insufficient information is available. This can be done by means of a food ethnography, which gives a descriptive analysis of the food system, food habits, and food culture of a population. In essence, the food ethnography is concerned with the question of how people deal with their food. When planning a nutrition or community health programme, it is essential to carry out a food ethnography. There is some evidence that, even in long-established nutrition programmes, community health programmes, and other forms of interventions, much basic information on how people deal with their food is lacking.

- To be used for more specific problem-oriented studies. Data on food habits and food consumption can be collected in conjunction with anthropometric, clinical, or biochemical assessment of the nutritional status of a population. Such data will provide an answer to the question of why a population is malnourished (overnutrition and undernutrition) and indicate ways to find a solution.
- Specific data on food habits and food consumption are needed during the planning, implementation, and evaluation of intervention programmes.

Socio-economic and cultural aspects of food and nutrition are also fields of interest in the social sciences. The study of food provides insight into the nature of a society on matters such as the social and economic structure, social differentiation, and culture. Eating is a biological, social, and cultural act. Scholars active in this field are nutritional anthropologists, geographers, and food-historians. Collaborative work between nutritionists or public health specialists and social scientists can be very helpful in understanding and solving nutritional issues.

The user of this manual should bear in mind that a survey only makes sense if its objectives are carefully defined: the purpose of the inquiry, why it is being undertaken, and what needs to be known. Suggestions and questions, questionnaires, and worksheets provided by the manual should be adapted to the specific needs and aims of the user. It is not possible to provide a list of standard questionnaires on food habits and consumption for every situation.

This manual is to be used as a guide and not as a set of fixed rules on food habits and food consumption. Sometimes the questions may be neither relevant for the user nor applicable for a specific situation. This manual is not meant to serve as an introduction for social or epidemiological surveys.

The manual comprises two parts: Part One provides insight into the dynamics of food habits and consumption and their socio-economic and cultural dimensions; Part Two provides practical information on small-scale studies to be carried out within the framework of a nutrition issue.

2. Food culture

2.1 Food culture: What is food?

Food is not just something to eat; it is an integral part of the culture of a community, region, or nation. Food is a relative concept. On a global level, humans eat everything that is not immediate toxic. However, when we take a close look at various distinct cultures, the situation is entirely different. What is considered edible in one culture may not be the case in another culture. Insects, for instance, are considered to be edible in several parts of Mexico, tropical Africa, and Southeast Asia. People in other parts of the world will be appalled by the idea of eating insects. The authors of the *Malawi Cook Book* rightly state that there is no reason for this. Insects are a good and cheap source of protein. Taxonomically, insects are not far removed from shrimp, which are considered a great delicacy in western societies (Shaxson *et al.*, 1979). Insects are not considered to be edible in Europe and most of the United States, despite recent attempts to introduce them to the diet. One of the reasons for this is a lack of sufficient edible insects in regions with temperate climates. In Mexico, by contrast, insects are packaged for sale in plastic sachets, cans, or jars. Also from a nutritional point of view, insects are good to eat, rich in proteins, fat, and the B vitamins (Bukkens, 1997; De Foliart, 1999; Van Huis, 2003). Generally speaking, people are likely to eat all sorts of dishes, provided they are not aware of the kind and origin of its components. At the very moment one realises the nature of the ingredients, a culturally determined aversion will develop.

People do not think of food in terms of energy and nutrients. What a community consumes is basically determined by four main interrelated basic factors:

- Geographical factors such as climate, soil conditions, lowlands, highlands, rural, urban, and the way how the available space for food production, food processing and transport has been organized.
- A time factor, socio-economic developments, and changes of a long and short nature; what we are used to eating or not eating is also influenced by the cultural legacy of previous generations.
- Culture will further determine attitudes toward food regarding what to eat and not to eat, and with whom, where, and when to eat.
- Taking the first three basic factors into account, a community's or household's access to food will further determine actual food intake.

It is important to realise that individual food behaviour may differ from what is generally accepted behaviour within a larger population. For example, meat is a much-appreciated food in the industrialized nations, but for various reasons it is rejected by some groups within the population.

This brings us to the question of: what are food habits? Food habits are the ways in which a community or a population group chooses, consumes, and makes use of available food in response to social, cultural, health, environmental, and economic pressures.

2.2 Food avoidances or food taboos

Food avoidances, often called food taboos, play an important role in many cultures when determining what food is and what is considered as edible. Food avoidance is a prohibition against consuming certain foods. The word "taboo" comes from Polynesian languages and means "sacred" or "forbidden", and has a quasi-magical or religious overtone. The term was introduced in the anthropological literature in the second half of the 19th century (Panoff and Perrin, 1973). In the field of food and nutrition, food avoidances are not necessarily connected with magical-religious practices, but are also associated with aversion due to unfamiliarity, culturally determined taste preferences, or health concepts. The more general term food avoidances is preferred in the field of nutrition.

It is of interest to note that food avoidance most frequently relates to animal meat, since in most cultures human beings have emotional relationships with the animals they have to kill to eat. An outspoken food avoidance of non-animal origin is the prohibition against alcohol for Muslims and some Christian denominations. A practice such as refraining from eating pork is not only a question of religious identity, but it also shows whether or not one belongs to a specific cultural community. In order to better understand the wide range of food avoidances from a nutritional point of view, it is useful to distinguish between permanent and temporary food avoidances (De Garine, 1972).

Permanent food avoidances

Foods that are permanently avoided are always prohibited for a specific group. The classic example of a permanent food taboo is the prohibition of pork practiced by Jews and Muslims. The Jewish prohibition against pork is found in the Book of Leviticus. Some anthropologists point out that food taboos are based on the failure of these foods to fit into the usual systems of classification. Foods that do not fit into these classifications are unsuitable for consumption, or unclean. According to the Koran, Muslims should not only avoid pork, but also blood, non-ritually slaughtered animals, cadavers, and alcohol (Benkheïra, 2000). In the case of both Jewish and Muslim avoidances, the foods themselves are considered unclean. A different concept of food avoidance is found in Hinduism. Hindus abstain from eating beef because cows are considered sacred.

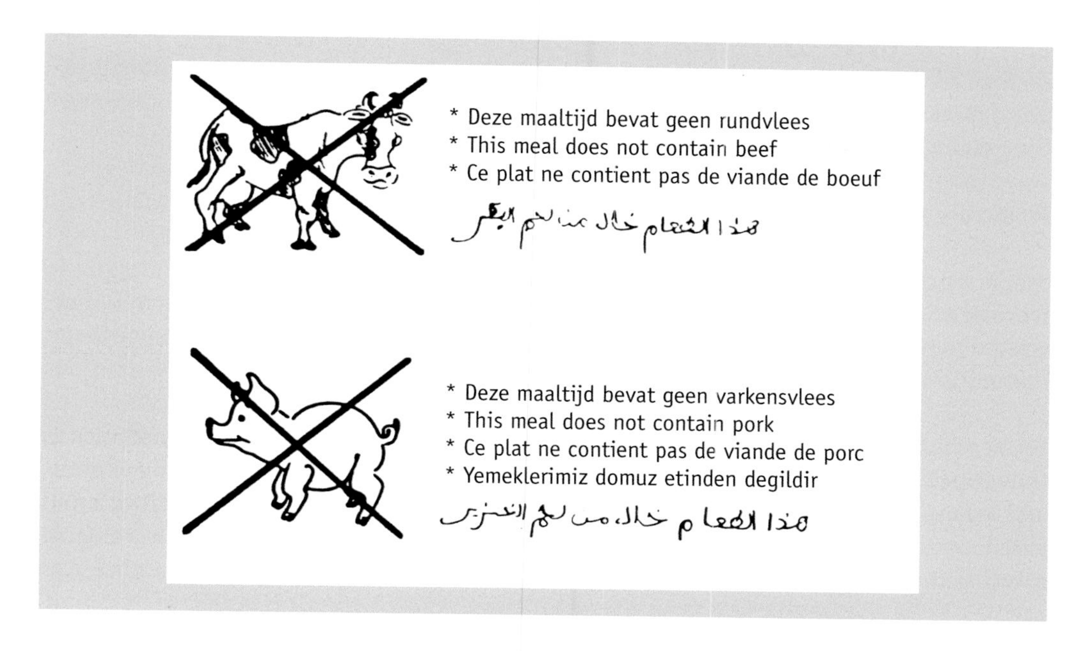

* Deze maaltijd bevat geen rundvlees
* This meal does not contain beef
* Ce plat ne contient pas de viande de boeuf
هذا الطعام خال من لحم البقر
* Deze maaltijd bevat geen varkensvlees
* This meal does not contain pork
* Ce plat ne contient pas de viande de porc
* Yemeklerimiz domuz etinden degildir
هذا الطعام خال من لحم الخنزير

Figure 2.1. Nutritional information on meals served in public institutions, permanent food avoidances (source: Netherlands Nutrition Center, 1985).

Various arguments have been used to explain the origins of such food avoidances, including religion, culture, and hygiene. Some scholars have rightly pointed out that there must be a logical and economic reason for rejecting certain foods (Harris, 1985; Simoons, 1994). The pig is an animal of sedentary farmers and unfit for a pastoral way of life. Herdsmen generally despise the lifestyle of sedentary farming communities. Communities based on or influenced by pastoral traditions refrain from eating pork.

Food avoidances are not part of the Christian way of thinking on what to eat and not to eat. However, Orthodox Christians in the Middle East as well as the Ethiopian Copts refrain from eating pork, referring to the Old Testament. Fasting, however, is practiced among Catholic and Orthodox Christians. Cats and dogs are not consumed in Western societies because of the emotional relationships developed with these pets. Pets are increasingly being "humanized" in such a way that eating them is seen as an act of anthropophagy or cannibalism. The feeling of closeness to certain animals can also be found in the savannah regions of West Africa, where certain West African clans consider dogs to be clan animals, based on the fact that they have been beneficial to the clan in the past. As clan animals, they are unfit for consumption. In antiquity, dogs were also considered unclean and unfit to eat. This is still the case in the Mediterranean area and the Middle East. By contrast, dog meat is popular in many parts of China, Northern Vietnam, and the mountainous regions of the Philippines (Cordero Fernando, 1976; Doling, 1996). From a nutritional point of view, dog meat is an excellent source of animal protein, and dogs do not require the grazing area demanded by cattle or other large ruminants.

Temporary food avoidances

Some foods are avoided for certain periods of time. These restrictions often apply to women and relate to the reproductive cycle. The times of temporary food avoidances related to particular periods of the life cycle include: pregnancy, birth, lactation, infancy, and initiation.

From a nutritional point of view, temporary food avoidances are of great importance as they concern certain vulnerable groups, such as pregnant women, breast-feeding women, and infants and children during periods of weaning and growth. Food regulations and avoidances during these periods often deprive the individual of nutritionally valuable foods such as meat, fish, eggs, or vegetables.

Pregnant women in a number of African countries avoid green vegetables, while fish may also be avoided. When asked why, women said the unborn child might develop a head with a fish shape appearance. Some of these avoidances may seem odd from a scientific point of view, but there is often an unnoticed logic behind them. In the first place, women are aware of the critical period of pregnancy and that much has to be done to ensure the successful delivery of a healthy child. Observing the rules of avoidance will give a pregnant woman the confidence in knowing that everything possible has been done for the benefit of the child. Nutritionists in Central Africa have observed that young children did not eat eggs. They were worried that a nutritious food was not being made available to this vulnerable group. The village elders gave a convincing explanation of why eggs should be avoided by children. They noted that in the past, the wise ancestors were much concerned about young children roaming around the villages searching for eggs and even chasing the brood hens away from their eggs. In order to avoid a depletion of the poultry stock, the elderly decided that eggs were harmful to young children and should be avoided.

A different form of temporary food avoidance involves the rules of fasting. In Christianity, for example, the most important periods of fasting are Fridays and Lent (the period from Ash Wednesday to Holy Saturday), during which time abstention from meat is observed. The Ethiopian Orthodox Church has a wide and complicated system of dietary rules and fasting (Shack, 1978). The Reformation in Europe largely broke with the tradition of fasting. In the Muslim world, Ramadan, the ninth month of the Muslim year, means strict fasting, even no drinking, from sunrise to sunset (Sakr, 1975).

Do food avoidances change and disappear?

Food avoidances may seem rather stable, but they are often under pressure because of changes in societies. Migration is an example of a powerful factor in the process of changing food culture. In Europe and North America, most Muslim migrants from the Maghreb, Middle East, and South Asia try to maintain their food habits, but some cannot fully resist the food culture of their new home country. A substantial number of Muslims begin drinking beer, wine, and even stronger spirits. Women tend to be less inclined to give up the avoidance of alcohol. The fear of pollution from pork often remains strong, however. In some European countries, Muslims refrain from eating in factory canteens out of fear that meals may be polluted with pork meat or fat. In contrast, many Jewish Europeans and Americans eat pork from time to time, or even on a regular basis.

Nutrition and health education has reduced temporary food avoidances among vulnerable groups in a great number of countries. In the humid tropical countries of Africa and Asia, where the raising of dairy animals is unfavorable, the rejection of milk as a food is diminishing. In the UK and other countries with Anglo-Saxon traditions, horsemeat is not part of the food culture. This in contrast to continental Europe, in particular France, where horsemeat is a well-known and appreciated food. It was considered to be unfit for consumption in Europe until the middle of the 19th century, but attitudes towards horsemeat changed dramatically. French pharmacists promoted the idea that horsemeat was suitable for consumption, and from a scientific point of view no threat at all to health. Discarded workhorses became a source of good and cheap meat for the growing working classes in urban France (Gade, 1976). The concept of horsemeat as food spread to other European countries, but not to the United Kingdom, where the horse remained a noble animal and the idea of eating horsemeat was received with disgust.

Dietary rules including food avoidances may be temporarily suspended during periods of emergency. The West African Fulani (Peul) pastoralists ordinarily avoid the consumption of fish (Simoons, 1994).During the dry season, the herdsmen have to move with their cattle from the northern savannahs to the south, near the Niger River. Because of the seasonal food shortage, herdsmen are more or less forced to turn to eating fish. In rural areas with a dry and a rainy season, people will collect the so-called "hungry foods" in the period of seasonal food shortage. Hungry foods are mainly wild foods, often not very attractive and tasty and, as such, normally avoided. They are consumed only as an emergency food.

2.3 Food classification systems

It is common practice in the nutritional sciences to classify food into certain food groups, such as carbohydrate-rich foods, protein-rich foods, foods of animal origin, or health promoting foods. Besides scientific classifications of foodstuffs, people in many societies have traditional systems of classification. In some Philippine communities, people classify their food into three categories: foodstuffs that allay one's hunger, such as rice; foodstuffs to satisfy an appetite (meat, green leafy vegetables); and taste (salt, peppers). Several cultures classify food according to a hot or cold classification system that looks irrelevant to an outsider, but is deeply seated in the understanding of the community. The hot-cold classification has nothing to do with the perception of temperature or taste. The aim of the classification system is to maintain or to restore physical health. The hot-cold classification can be found in a wide range of countries, including some in Central and South America, India, Sri Lanka, and several parts of China. This system is closely related to the theory of the four humours described by the Greek classical physician and philosopher Hippocrates. The key element of system is that foods consumed should be in balance with each other. A so-called hot food should be brought into balance by also eating a cold food. If this is not done, a disease will occur. On the other hand, a disease can be cured by taking the hot-cold classification into account. The reason why a certain food is either hot or cold is difficult to understand by an outsider. In Central Asia, for example, food can be either hot or cold for most ethnic groups, serving both medicinal and nutritional functions.

A typical meal in rural Mexico may start with rice (cold), followed by a soup of hot and cold ingredients, and end with dark beans (hot). This dichotomy of hot and cold foods has its origin in classical Spanish medicine and in the indigenous cultures that preceded the Spanish conquest (Molony, 1975). Some groups in Western society have been inspired by the old Chinese dichotomy that most foodstuffs have a *yin* or *yang* nature.

Food can also be classified according to a religious basis as well. The Koran classifies foods as lawful (*halal*) or unlawful (*haram*). *Halal* signifies food that is accepted in the eyes of God. Most animals are *halal* (not the pig), however, to be *halal*, meat must come from animals slaughtered ritually (Benkheïra, 2000). Additional aspects of food avoidance will be further discussed in section 2.6. It is not within the scope of this manual to go further into details of Islamic laws regarding food. In Judaism, food is primarily defined by dietary laws. Under the classifying terminology of *kosher* versus non-*kosher*, the system of dietary prescriptions primarily involves the consumption of animal food. The logic behind the system is well presented by Mary Douglas in her classical study *Purity and Danger* (1966).

The food classification system can be very complicated and researchers are strongly advised to get adequate information and help from local advisers.

2.4 Social role of food in society

Food has basically two major roles in a society:
- The first is a health role: the nutritional health of population groups or a community is a condition for development, maintaining social justice, and stability.
- Secondly, food performs a number of interrelated social roles in a society.

The various social roles of food in a society has an effect on the way how people make use of available food. The following social roles of food can be distinguished:
- gastronomic meaning;
- means of expressing cultural identity;
- religious and magical meaning;
- means of communication;
- expression of status and distinction;
- means of influence and power;
- means of exchange.

2.5 Gastronomic meaning

It is not always fully realized by professionals working in the field of food and nutrition or policy and planning, but people eat for the pleasure of it. The organoleptic properties of a food have an influence on whether people accept or reject a food or dish. The pleasantness of a food, food product or a dish is determined by variables such as taste, odour, temperature, appearance, structure or texture. The pleasure obtained from food has both a psychological and a cultural basis. The taste and appearance of food differ from region to region, and among socio-economic groups within the same society. In Europe, there is, generally speaking, a preference for more soft types of food. In Tropical Africa, people often like to chew foods, such as meat, as chewing is thought to give better satisfaction when enjoying the taste of the meat. In many rice-consuming countries of Asia, there is an outspoken preference for the granular structure of boiled rice. In other parts of the world, such as in Italy, rice (risotto) has to be a bit sticky. Glutinous staples are highly appreciated in tropical Africa.

Gastronomy, a term now widely used in many countries, means the practice and art of eating and drinking well. The Frenchman Brillat-Savarin stated in his classical book entitled *The Physiology of Taste* (1825/1984), that animals just take their food, but man eats and has developed a philosophy about his food, centred around gastronomy. Eating well and having access to a wide range of sophisticated foods has been the privilege of the higher classes in most societies, while farmers and poor urban dwellers have to be content with simple food. The high cuisine in many cultures originates from earlier ways of life at feudal or royal courts (Goody, 1982). The oldest written records of a highly developed cuisine (at royal courts) has to be found in Mesopotamia (Iraq), nearly 4000 years ago (Bottéro, 2002). The geographical dimensions of food, the process of diffusion of (sophisticated) food, and dishes among larger parts of the society will be discussed in Chapter 3.

Consumers in industrialized societies sometimes complain that food is becoming less tastier, both fresh and processed. The further expanding chain between the urban consumers and the food producers is an enormous challenge to the food industry. The demands on food are high. The food industry must offer safe foods that retain organoleptic qualities and nutrients during
. Developing countries with a recent food industry and an interest in processing national food products and dishes for the modern consumer are in fierce competition with imported foods. The *Slow Food* movement (versus fast food industry) in Europe started as a counter reaction against the effects of globalization and levelling out of local food culture by promoting traditional regional products and the authenticity of taste. Slow Food started in 1986 in Italy and consists of a vast European network of quality food producers and sympathisers.

2.6 Means of cultural identity

Food often provides elements for the cultural identity of a group of people, community, or a nation. From this perspective, people can be rather emotional towards their national food. Rejecting the food from a community or a country is not only felt as an offence, but is also seen as the rejection of an entire culture. Among traditional farming communities in Mexico, maize is identified with life and attitudes toward it are often religious. Other staple foods such as rice or yam have a role as a means of identity, such as in West Africa and the well-known Yam Festivals.

In some cases, outsiders may identify a community or a nation with a food, often in a prejudicial nature. The Inuit artic population groups, which means people, were until recently referred to as Eskimos. The name Eskimo, meaning eaters of raw meat, was once given to them by neighbouring Algonquin Indians, referring to their particular food habits (Panoff and Perrin, 1973). The Dutch are sometimes nicknamed as cheese heads, referring to their national food, cheese. In Danish food culture, pork is of essential importance and has become a symbol of economic success and part of Danish identity (Delavigne, 2002).

As a counter movement against the process of globalization and the danger of losing regional and national identity, interest in local cuisine is rising. This is revealed in many countries by a flow of books and articles in magazines and newspapers on national or regional cuisine. National dishes now also appear on the menu lists of restaurants

catering mainly for visitors. It is important to realise that so-called national dishes are often originally regionally based dishes. A national cuisine is based on a combination of a range of regional kitchens.
In the industrialized societies, new emotional feelings towards food emerged. The "health food" shops, selling *natural* or *organic* products (foods grown without artificial fertilizers, insecticides or foods containing no additives such as artificial colouring, preservatives, *etc.*) have found a place in the industrialized society. The followers of *macrobiotic* diets and vegetarians identify themselves with a diet based on health and ethical grounds. It can be considered as protest against the industrial society with its over-consumption, a food economy based on waste detrimental to the environment.

The strict use of food creates and maintains boundaries and common identity between population groups. This is particularly the case when dealing with food avoidances. In Muslim countries with Christian communities or the other way round in the European Union with Muslim Communities, the eating or non-eating of pork clearly distinguishes between the two different groups (Benkheïra, 2002). The same applied until recently in the European Union, where the eating of fish on Friday indicted whether people were Catholic, Orthodox, or Protestant.

2.7 Religious or magic meaning

There is much religious and magic symbolism associated with food and this should be analysed within the context of the community and society (Kilara and Lya, 1992). The role of food in religion should be taken into account in nutrition intervention programmes such as nutrition education or food relief. The attitude of people towards their staple food has a sacred character in many communities, and dietary regulations about food are used in God's services. Some examples will be discussed. Food in Islam is considered to be a gift of god. Before the meal is started in Middle Eastern households the word *Bismillah* (in the name of God) is uttered by all. When finished, one says, "To God be thanks" (Roden, 2000). Praying and thanking God for a meal and the blessing of food is practiced in many Christian communities. Bread is considered as the body of Christ and wine is his blood during the Communion; in Orthodox communities, all sorts of symbols are marked on the bread. In countries and regions based or influenced by Hinduism, such as on Bali, rituals are performed to satisfy the rice goddess Dewi Sri. A thanksgiving festival is dedicated to Dewi Sri just before or after the rice harvest, which involves the preparation of sumptuous offerings of food. In the Savannah regions of tropical Africa, the *chef de terre* or chief of the earth, performs all sorts of rituals before the preparation of the fields for planting can take a start (De Garine and Koppert, 1988). Of a quite different nature is the use of food among some groups in magic rituals. Food can be used in such a way that it can exercise power on other persons.

2.8 Food as means of communication

Food offered to a visitor or guest may put people at ease and facilitate communication. Food also plays a significant communication role in a community. Hospitality is a striking social institution in many traditional societies. Hospitality may even go so far that it becomes a heavy burden on the household budget. The exchange of gifts and goods at social occasions has a strongly competitive element. This element also prevails at many dinner parties, weddings and other festivals in modern urbanized society. In households in urbanized communities where both husband and wife work outside the home, the evening meal is often the only occasion where the whole family can be together. Offering food to the ancestors is a method of keeping in contact with them. Offering a libation to the ancestors before drinking is most common in African societies.

The *slametan*, a ceremonial meal in Indonesia and in particular in Java and Madura, is an interesting example of sharing food and signifying bonds between people. The *slametan* is a ceremonial meal held at specific occasions such as births and marriage, but also in modern settings such as for the opening of a factory or office. The word *slametan* is derived from *selamet*, which means being well, safe, blessed, or prosperous. Among the Javanese, well-being is very much linked with harmony. One has to live in harmony with one's fellow men, with God, and with the ancestors. Harmony can be restored, strengthened, and enhanced by organizing a *slametan*. The food consumed at a *slametan* is high in quality in comparison to the daily menu. Animal foods in particular are not, or to only a very limited degree, a part of the daily menu of ordinary villagers. The guests at a *slametan* are men and count on a good meal. Although the men have the best part of the meal, women and children are not left out. People believe that neglecting to give a *slametan* will bring bad luck, even more so than when committing theft. On the other hand, people often have to borrow and accumulate debt in order to be able to give a *slametan*. Food sharing rituals strengthen social bonds and, in the case of the *slametan*, achieve and maintain harmony (Niehof, 2001).

2.9 Food as expression of status and distinction

Food is a sign of wealth and status. The French sociologist Pierre Bourdieu (1989) has pointed out that how people choose and consume their food is also a means of becoming distinct from other population groups of society. It is the same manner by which people choose their clothes and get dressed or furnish their homes. Another aspect is that the effects of food on the body also serve as a means of distinction. The prevailing concepts on what is a healthy and beautiful body differ within and between various cultures. In societies where a corpulent body is traditionally appreciated, health education programmes against obesity are more difficult to implement compared to societies where a slim figure is considered as ideal. The higher status of overweight-looking people is diminishing in emerging economies.

Food can play a role in social snobbery, something which is to be found to various degrees in most societies. All societies have prestige foods, mainly reserved for special occasions. Foods of animal origin are very prestigious. As a result of the rapid process of globalization, processed foods from industrialized countries are increasingly imported by developing countries. Multinationals have established food factories in several of these countries. In the less industrialized countries, these processed foods have a high prestige. A competition is going on between the industrial processed foods, both imported and locally produced, and the traditional foods manufactured by small-scale enterprises (Trèche *et al.*, 2002).

2.10 Means of influence and power

Food can be used at several levels as a means to exercise influence and power. Those people or groups in control of the food supply and distribution can also control society. This has become very clear in countries suffering from civil unrest and war. Food is utilized to manipulate the situation, to favour allies and to withhold food from opponents. The former Yugoslavia of the early 1990's is a bitter example of how the combating parties utilized food as a weapon by besieging civilian communities and blocking food transport. Despite its humanitarian nature, bilateral food aid can also be used to influence governments or population groups of recipient countries. Food at household level food can also be used to gain influence by those responsible for the family food stocks. Parents may reward their children with giving some special food or punish by withholding it.

2.11 Means of exchange

Food and food products in rural societies are often used as a means of exchange to get other foods or non-food items. Pastoral populations exchange livestock such as cattle or sheep with cereals from sedentary farming communities (Little *et al.*, 2001). In farming communities it is not unusual to pay partly or wholly in kind (food) when buying farm equipment or utensils from other villagers. The dowry in many cultures also comprises food and food products.

3. Geographical dimensions of food

3.1 Dietary patterns and their links with geographical zones

Dietary patterns are often linked with geographical zones:

- Food consumption and food availability data indicate a relationship between what a community eats and the existing dominant food system of a particular area. Domestication of plants in response to climate, soil, and topography has led to the development of two major cropping systems (Harris, 1969; Anonymous, 1985).
- Seed agriculture, primarily dependent upon crop plants reproduced by seeds, nearly all belonging to the grass family (*Gramineae*).
- Root-and-tuber culture, dependant mainly on vegetative reproduction.

Seed agriculture, especially of sorghum and millet, is the indigenous mode of agriculture in the drier tropics and sub-tropics. Wheat is the staple crop of the dryer and temperate climate zones. Rice dominates the food system of the tropical and warm temperate zones of Asia. Maize is the indigenous cereal of the more humid tropical areas of the American continent. Root-and-tuber agriculture, primarily cassava, yam, and taro, is most highly developed as an indigenous agriculture in the humid tropical lowlands of America, Southeast Asia, and Africa. The root-and-tuber agriculture of the cool climate of the Andean highlands is based on potatoes and some other minor root crops. The West African region is a clear example of the relationship between geographical zones and major dietary staple foods (Figure 3.1).

Plant and animal domestication started some 10,000 years ago, leading to the ability to feed larger settled communities and, ultimately, to the amenities of urban life. The control of cultivating and producing sufficient food both in quantity and variety has been of fundamental importance in the development of mankind (Diamond, 1999). However, the collection of wild foods in various rural areas along roadsides, fields, and forests is still a non-negligible activity from a nutrition point of view (Kuhnlein, 2003). Even in industrialized countries such as South Korea, wild/gathered foods are much appreciated for their taste and are important as a source of cultural identity (Pemberton, 2002). Two main agricultural systems exist in Southeast Asia: the dominant *sawah*, or wet-rice cultivation, and swidden cultivation (shifting cultivation). Generally speaking, rural communities in many parts of the world have a dietary pattern corresponding to the prevailing food production system of the geographical zone where they are living, but these links are usually less marked in the cities.

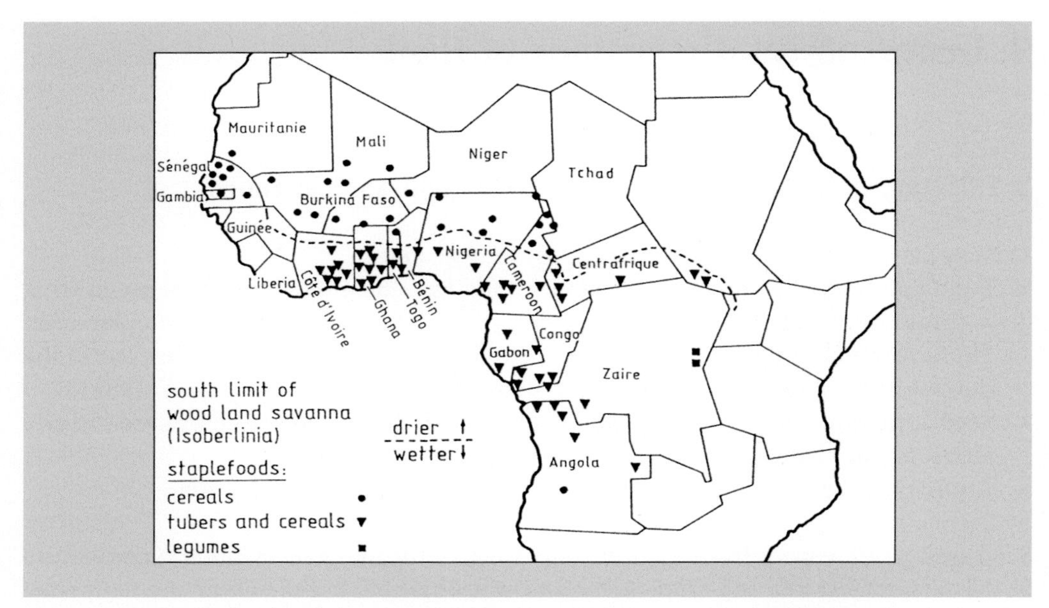

Figure 3.1. Staple foods of the two major geographical zones of West Africa: dry (one rainy season) and humid zones (two rainy seasons) (adapted from Watier 1982).

The geographical dimensions of food have never been static, and food diffusion has been continuous over time. Foods from one continent have been diffused and accepted in other parts of the world. An example would be the diffusion of maize from the American continent to tropical Africa, which was already underway in the 16th century. The process of food diffusion will be discussed in Chapter 3.3.

In those parts of the world where rural communities mainly consume self-produced food, the situation differs from the towns and cities. From a dietary point of view, they can be considered as food enclaves with a diversified diet situated in a rather monotonous food zone, as was also the case in pre-industrial Europe (Pitte, 1991). Considering the geographical dimensions of food, cultural factors should be taken into account to avoid a deterministic approach. The olive tree and the grape (wine growing) are characteristic of the Mediterranean regions. The olive is cultivated and consumed wherever it is technically possible, while wine growing is confined to the northern part of the Mediterranean. In the Islamic south, grapes are only cultivated on a limited scale, as drinking wine is not permitted (Lignon-Darmaillac, 2001).

Dependence on a single food crop such as rice, maize, cassava, or plantain, as is often the case in poor rural areas, can often lead to nutritional deficiencies. In the forest zones of tropical Africa where a root-and-tuber food culture is practised, the diet is characterized by a deficiency in proteins and some B vitamins (FAO, 1990a). Vitamin A and iron deficiencies are common in arid areas, for example, particularly during the long dry season when vegetables are scarce and animal production is limited by an absence of green pasture and other animal fodder.

3.2 Dairy and the geographical context

The geographic dimensions of food can be well illustrated by the appreciation of milk as a food and its prominence in the diet. Looking at contemporary food and consumption habits, a distinction can be made between populations with a long-standing dairy tradition and populations where the use of milk was absent until recently. Is there an insuperable prejudice against milk among non-dairying populations? Some scholars like Marvin Harris (1985) went even so far as to divide the world into lactophile (milk loving) and lactophobe (milk hating) populations, the later to be found in China, Southeast Asia, humid tropical Africa, and the indigenous populations of the Pacific and the American continent (see Figure 3.2).

The use of milk as a food probably originated in the fertile crescent of the Middle East, one of the cradles of plant and animal domestication. It started some 6,000 or 7,000 years ago with the domestication of ruminants such as goats, sheep, and cows. The skills and techniques of dairying spread from this region along the Mediterranean to Europe, North and East Africa (particularly in the highlands), and the savannas of West Africa.

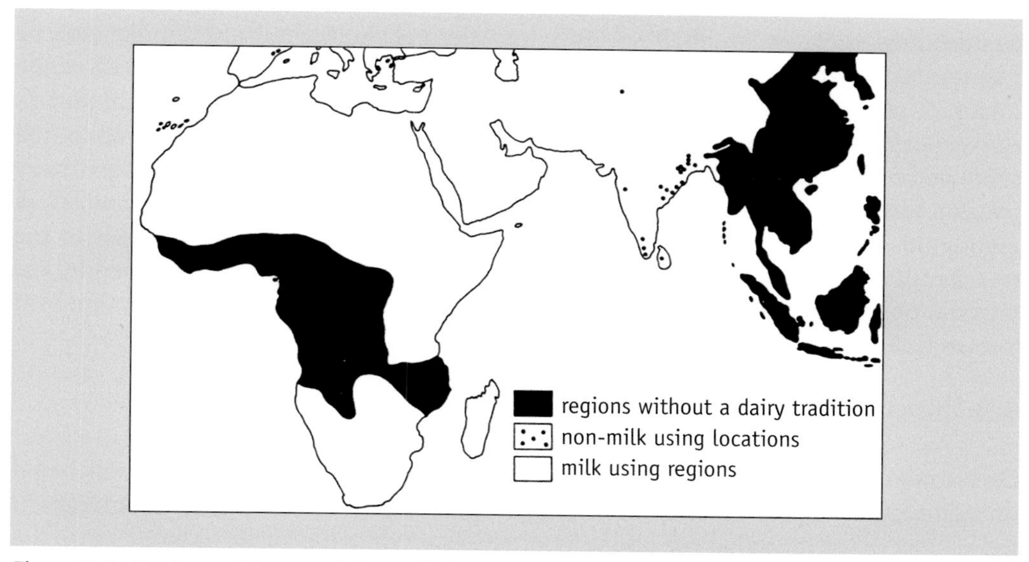

Figure 3.2. Regions without a dairy tradition (adapted from Simoons, 1973).

Another diffusion took place to the Indian sub-continent and Central Asia. In dairying, a distinction can be made between the cattle-keeping pastoral peoples and the sedentary cattle-keeping farming communities.

In Central Asia, fermented milk products are important in the cuisine of the Turkic speaking populations, while yoghurt and fresh cheese are commonly used in the cuisine of the Middle East and India. In addition, ghee, a clarified butter, is highly appreciated in the Indian sub-continent, where the cow is very important as a sacred animal and supplier of milk in the Hindu culture. Generally speaking, milk and milk products provide a major contribution to the diet of the pastoralists. For example, milk is not only a culturally preferred food among East African pastoralists, but is also a dietary staple contributing up to 90% of the energy intake in some parts of the year. Cattle are seen as milk producers rather than as a source of meat or as an exchange item for grains (Little *et al.*, 2001). Fresh milk quickly sours under warm conditions because of fermentation. Until recently, fermented milk was not widely used in the temperate regions of Europe except for buttermilk in dairy countries such as the Netherlands.

Generally speaking, dairying did not reach the humid tropics of Africa and Southeast Asia, where the climatological conditions are unfavourable for keeping cattle. The reason why dairy did not develop as a major activity in China, Korea and Japan are less clear. Because of the absence of cattle, dairying and milk use were unknown in pre-Columbian America. The indigenous livestock of the civilizations of the Andes, the lama and the alpaca, were used as pack animals and for their wool and meat.

Besides cow milk, other kinds of milk and derived milk products should also be mentioned. Second to cow milk is the production of buffalo milk, of which 88% of the total production is produced in the Indian sub-continent. Compared to cow and buffalo milk, sheep and goat milk are of lesser importance. In many local communities of the semi-arid zones, however, sheep- and goat milk are significant components of the dietary pattern. It is of interest to note that the total production and consumption of camel milk among the pastoral peoples in the arid zones of Africa and Asia is comparable to the total production of sheep and goat milk (FAO, 1995). Mare's milk is consumed by the pastoral populations in Central Asia, where it is fermented and consumed in the form known as *kymus*, a mild alcoholic drink.

3.3 The changing place of dairying

Do the populations in regions without a dairy tradition have a revulsion against milk based on lactose intolerance? Primary lactose intolerance results from an apparent decrease in the activity of the intestinal enzyme lactase, which provides the ability to break down the lactose or milk sugar, which is a prerequisite for a good absorption in the intestinal tract.

Lactose intolerance can occur between the ages of two and five years. This condition is present in various degrees in about 75% of the world population. Most likely this can be ascribed to the fact that these populations do not use milk after infancy.

Results from studies carried out on lactose intolerance suggest a genetic basis for primary lactose intolerance among those with no milk-drinking tradition. A late evolutionary development among dairying populations probably created a tolerance by the use of a lactose rich diet in the form of cow milk over many generations. Small but nevertheless significant quantities of milk taken throughout the day can be tolerated by older children and adults, as well as by ethnic groups known to be lactose intolerant (Scrimshaw and Murray, 1988).

It should be noted that it is not easy to interpret symptoms of discomfort associated with milk drinking. Intestinal infections, which often result in malnutrition, may sometimes cause temporary secondary lactose intolerance. Congenital lactose intolerance caused by an absence of intestinal lactase is very rare. This occurs immediately after birth and has very serious consequences for the infant. The use of dairy foods has extended in most parts of the world despite the occurrence of lactose intolerance.

Traditional milk-using populations

Milk and dairy products are consumed almost exclusively by the middle and higher socio-economic classes in the cities and regions with a dairy tradition. Dairy is much appreciated but not easily accessible for slum dwellers. In the rural areas of India, for example, improved transportation systems and cooling have brought milk within the reach of the urban middle class. However, milk has remained beyond the reach of the rural poor. Small milk producers sell most of their milk, so hardly anything is left for household consumption. The cattle-keeping nomads in West Africa are threatened by the increasing expansion of the cultivated areas of the farming communities. As a result, nomadic communities are often compelled to adopt a sedentary way of life, resulting in a diminishing share of milk and milk products in their diet.

Populations without a dairy tradition

Milk and milk products have obtained a place, although often a modest, in the diet in regions without a dairy tradition. This is, however, an urban middle class phenomenon. Indonesia, in particular the island of Java, is an interesting example of how dairy was introduced first in the colonial era and later by the activities of multinational firms (Den Hartog, 2002). The government of Indonesia required foreign dairy firms to set up a local condensed milk industry in the 1950's and 1960's, using both local and imported raw materials. Tinned milk products, particularly sweetened condensed milk, are more popular than fresh milk, which is an expensive beverage and highly perishable. Tinned milk products have found their way not only into the urban and peri-urban markets, but also into the *tokos* (shops) and *warungs* (stalls) of the rural areas.

Nowadays milk is used not only as a beverage, but also as an ingredient in ice cream or tonics. Advertisements, for example, suggest that sweetened condensed milk can be poured over green beans or over a black glutinous rice porridge. The coming of UHT milk (ultra high temperature treatment) in carton containers has also had an impact. Long life milk in small handy carton containers with various flavours can be found in all major cities.

No indications of a fundamental and insurmountable prejudice against milk and dairying as a whole could be found in Indonesia or other traditional non-dairying countries. The price, rather than unfamiliarity, remained an obstacle for its popularity. It seems that the occurrence of primary lactose intolerance does not deter a limited adoption of milk in the dietary pattern.

3.4 Food diffusion and globalization

The way people and communities choose and consume certain foods has never been static. Food habits are changing constantly, for better or worse, by external influence or by developments from within the society itself. Some attention was given to the dynamics of food habits in previous sections. Globalization is in essence not a new phenomenon. It was seen that food crops and animals diffused throughout the world in pre-historic times.

The diffusion of food from one continent to another and from one society to another has been extensive in the past 400 years, starting with the European expansion in the 16th century. The process became intensified in the 20th century with the emergence of large multinational food firms. Trade, war, and migration have been powerful factors in changing food habits over the course of the time (Fumey and Etcheverria, 2004). As a result of the European expansion, potatoes, sweet potatoes, cassava, cacao, tomatoes, lima beans, groundnuts, and turkeys went from the Americas to Europe, Africa, and Asia. Foodstuffs such as rice, tea, sugarcane, and many fruits spread from Southeast Asia to other tropical and subtropical areas of the world. Cattle, sheep, and goats went from Europe to the Americas.

These foods are so well established in some countries that people consider them to be indigenous. Cassava and maize are now so well incorporated into the cuisine and diet in many parts of tropical Africa that these foods have become a part of the prevailing food culture. The wide range of varieties of the fermented African maize porridges is an interesting example of food innovation (Adandé, 1984). The fermented maize porridge with its complicated preparation is far removed from the original maize type of flatbread, the *tortilla* of Mexico and Central America. Parallel developments occurred in Europe with the introduction of the potato. One of the first scholars dealing with food anthropology, Margaret Mead (1960), once remarked we should not only ask how do we change food habits, but how do food habits change. Indeed, desirable food and nutrition policy changes can only be introduced with a knowledge of existing food habit trends, interrelations, and dietary patterns. Processes of changing food habits and dietary patterns are complex and often inscrutable. To obtain a better insight into the nature of food in change, a distinction should be made between autonomous and induced changes.

Autonomous food-related changes

Food habits and the food pattern of population groups or a community are closely associated with the society and country of which they are a part. Thus, the socio-economic dynamics of a society and country will have an impact on food production and food habits and, ultimately, on food consumption and health.

Five main factors will impact food habits: population growth, urbanization, rural transition, a rise in the standard of living, and education. Several aspects of these factors will be discussed in the next chapters. The term "autonomous food-related changes" is used because these main factors act independently from food and nutrition policy and nutrition interventions.

Inducing food changes

Food habits are also directly influenced by a number of actors, each with its own specific objectives, target groups and basic assumptions. These actors are mainly governmental institutions, NGOs, and the food industry. These actors try to influence and change the food habits of their target groups by means of marketing or nutritional and health education.

The governmental institutions (*e.g.* Departments of Health, Agriculture, Education, Nutrition Programmes) aim to promote good food habits for better nutrition by means of education and extension services. The private food- and nutrition-related institutions of both national and international Non-Governmental Organizations, or NGOs, give additional support in areas where government funding is insufficient or lacking. On the question of what is indeed good to eat from a nutrition and health point of view, governmental institutions can receive technical backstopping from national food and nutrition institutes and international organizations such as WHO, FAO and UNICEF.

Of a complete different nature and opposite point of view are the marketing activities of the food industry. Food marketing aims to change food habits and purchasing behaviour so that certain kinds of food will be bought and consumed. The objective of food marketing is to promote and sell certain foods, and the health implications are considered as a derivative. Health and nutrition education are often at odds with food marketing. In countries with well- established administrations, government and consumer organizations try to set up a dialogue with the food industry and to set out a code on food promotion on how to deal with health claims in radio and TV commercials and newspaper advertisements. Compared to the food industry, nutrition education is always at a disadvantage, as it is handicapped by limited resources and often has to deal with a wider range of foods than food marketing practises. The nutritional message is often less appealing than promoting biscuits, beer, or soft drinks. There are, however, good examples of collaboration between the private food industry and governments, such as the iodization of salt and vitamin A and D fortification.

What does "changing" mean?

It is necessary to specify what is meant by "change" when discussing the concept of changing food habits. The concept of changing food habits comprises the following elements:

- *Changes in the use of already known foods.* This can be the increase or even decrease in the use of known foods such as meat, dairy, or staple foods. The consumption of traditional foods is often declining. On a general basis, people tend to consume more food of animal origin, fats, fruits and vegetables and less starchy staple foods as a result of increases in standards of living. It also means a modification or a sophistication of already known foods. Examples of this are industrially prepared local dishes, fruit juices in bottles, and the substitution of traditionally brewed beers in tropical Africa by industrialized kinds of beer in bottles.
- *New foods previously unknown in the food pattern.* In the past, new foods have replaced traditional staple foods in several regions of tropical Africa, such as sorghum and millet by maize and cassava, for example. Of a more recent nature is the process in Papua New Guinea whereby the staples such as yam, cassava, and bananas are increasingly being replaced by white rice and bread (Hodge *et al.,* 1996).
- *Changing attitude towards a food or a category of food.* In various societies and among population groups, one can observe a rise or fall in the appreciation for certain foods. Imported food products are often highly appreciated. The traditional appreciation for coarse foodstuffs is diminishing. People in urban areas prefer the white types of rice and white bread. Even in countries with a strong culinary tradition such as India and China, western kinds of fast food and pizza have obtained appreciation among young middle class consumers.

3.5 Acceptance of new foods

The diffusion and acceptance of new foods takes place at two levels. The first level is the geographic or horizontal diffusion of food or food products, from one continent or region to another, as discussed in the previous section. Second is the vertical diffusion in the recipient society or country, which is the acceptance of new food among socio-economic classes or ethnic groups.

Why have previously unknown foods been accepted in the food pattern? Food diffusion and acceptance is complex. For a better understanding of the process of accepting new foods, four major interrelated factors will be discussed:

- In the first place it depends on the nature of the food: is it a staple, a condiment or a drink? Historically, the search for condiments such as black pepper, cloves, and cinnamon resulted in an improvement of the taste of the diet in Europe. It also had far reaching consequences for the relations between Europe and Asia in modern history. The same can be said of man's desire for things that are sweet: sugar, once a rare commodity, is now accessible for most people in the world. The establishment of plantations in the Caribbean and Central America in the 17th century based on slave labour caused the African Diaspora (Mintz, 1985). Tea from China and coffee from Ethiopia have become one of the most important beverages of the world, ensuring a safe way of drinking because water has to be boiled as part of the preparation (Mauro, 1991).
- The degree to which a new food can be incorporated into the local food production system. Would it be possible to cultivate the new food locally or must it always be imported? Wheat, in the form of bread or biscuits, is now increasingly consumed in tropical countries, but wheat cannot be cultivated in the humid tropics because of climatological reasons. This is one of the reasons why wheat did not become a dietary staple in these geographical zones. However, when the new food can fit into the local farming system, it may have a considerable impact on the recipient society. The staple foods cassava and maize are interesting examples. Cassava, with its high yield of energy, has on many occasions saved populations from famine, despite its low protein content (FAO, 1990a). In Indonesia, in areas with poor soils on the island of Java, cassava was already introduced in the 19th century because rice yields were insufficient to ensure household food security. The same occurred even in an earlier stage in the humid tropics of Africa. Maize was accepted in many tropical countries for similar reasons (FAO, 1992). In Europe, the introduction of the potato became successful when wheat production became insufficient to feed a rising population in the pre-industrial period. Two or even three times more people could be fed from a field of potatoes when compared with wheat.

- Food culture is an important factor in the process of accepting a new food. A new food is more easily accepted when it fits into the prevailing concept of what is edible and when the local culinary techniques can be applied. The duration of the food preparation time also plays a role. There is a need for fuel saving foods because of firewood shortages. In urban areas of tropical Africa, bread prepared from cheap imported wheat has become a popular food at breakfast or as a snack. Bread is ready to eat and it saves time and fuel. Taste and prestige are also important reasons for the acceptance of new foods, price in relation to purchasing power being the limited factor.
- The role and power of the actors further determine the way and degree of food acceptance. In the past, colonial powers played a crucial role in food diffusion. Boarding schools in Benin during colonial times played a major role in making the future elite acquainted with French foods. After independence, truck drivers brought tinned sardines, bread and red wine to the northern part of the country (Cornevin, 1981). The contemporary food industry, both multinational and national, is a mighty actor in the process of food diffusion.

New foods are not always readily accepted among recipient populations and a resistance to change may arise when the advantages are not clear or absent. The soybean in Asian countries such as China, Japan, and Indonesia is an age-old food crop with a wide range of preparations and applications. The United States is currently the biggest exporter of soybeans. The soy export industry has made various efforts to have soybean accepted into the diet in tropical Africa, which has stimulated local soy production. One of the arguments for its introduction is to contribute towards a reduction of protein deficiencies in the diet. The soybean has received much resistance nevertheless. People often disliked the taste and soybean preparations needed more cooking time compared to indigenous beans, hence requiring more wood fuel.

Food innovation

The acceptance of a new food in the diet of a population can be considered as a food innovation. Two basic types of food innovation can be distinguished: an expensive innovation and an emergency innovation. Expensive food innovations are first adopted by the upper classes of a society. Through close contacts, and considering the upper class as a point of reference, foods often move down the social scale to the middle and lower classes. This process will be stronger in periods of rising standards of living with improved production and mass transport techniques. Finally, the food will lose its exclusive character and become part of the diet of the lower classes. Soft drinks such as the colas were originally so expensive in most countries that only high-income groups could afford them. Drinking cola or bottled beer became a sign of being modern. Soft drinks gradually became accessible to the lower income groups and lost their prestige.

As a result of continuing marketing practises, soft drinks have become an integral part of the dietary pattern.

On the other hand, food innovations may also occur in periods of economic distress. It was mentioned in a previous section that cassava and maize were accepted out of emergency need, as the production of traditional foods could no longer ensure food security. These foods were introduced to the diet of the middle and higher classes from the lower classes. Cassava is now fully accepted in the diet of all classes in Benin, but in the beginning of the 20th century it was looked upon as food of the lower classes (Adandé, 1984). Whisky is another example. Once a drink of poor Irish and Scottish farmers, it is now a prestigious drink in many countries. Food diffusion and the incorporation of new foods into the diet is an everlasting process. The question of why urbanites are so receptive to new foods will be dealt with at consumer level in the first section of Chapter 4.

4. Nutrition in transition

4.1 Urbanization: nutrition in transition

Urban developments

The world as a whole is undergoing urbanization. In 1950, 29.7% of the world population was living in towns and cities; in 2000 it was already 47% and is expected to increase to 60.3% in 2030. Table 4.1 indicates that this process of urbanization is much faster in the less developed regions. The transformation of rural settlements into cities, rural-urban migration, and population growth as such are major determinant consequences of high urban population growth in the less developed regions over the next thirty years. In combination with the universal reduction of fertility levels that is expected to occur in the future, these changes will lead to a reduction of rural populations. The expected decline of rural populations by 2020 is similar to what has happened with rural populations in more developed regions since 1950 (UN, 2000).

From a food and nutrition point of view, this process of urbanization implies a number of questions to be answered and points of concern.

- How will an increasing non-food producing population be fed? This is a major issue for food policy makers. This means placing more emphasis on planning the urban food supply and food distribution system to ensure food security and food safety for low-income groups.

Table 4.1. Urbanization by region expressed in percentage of the total population, and the world population in billions 1950-2030.

	1950 %	1975 %	2000 %	2030 %
World	29.7	37.9	47.0	60.3
More developed regions	54.9	70.0	76.0	83.5
Less developed regions	17.8	26.8	39.9	56.2
World population in billions	2.52	4.07	6.06	8.11

Source: UN Population Division, 2000

- How will the quality of the diet be maintained and improved? This is basically a food and health issue. Even when the urban diet meets nutritional requirements, environmental issues have to be taken into account. The latter concerns sanitation, safe water supply, effective sewerage systems and refuse collection. City councils should be involved with setting up and implementing urban food and nutrition policy. Such a policy should obviously be in line with national food and nutrition policies but adapted to the specific local urban situation.
- In which direction is the urban trend, towards less access to food for low-income population groups or leading towards an urban affluent society? At present, the phenomenon of the double burden of nutrition emerges in many cities. It relates to two forms of malnutrition: undernutrition and related deficiency diseases among the urban slum dwellers, and the various forms of overnutrition, in particular obesity, to be found among the middle and higher classes.
- In which way is the local food industry in a position to produce food products suitable for the urban way of life, with special reference to the low-income consumer? From a nutritional and local employment point of view, the small scale and artisan type of food industry should receive support from the city councils (Trèche *et al.*, 2002).

Two distinct trends in the process of food system urbanization and globalization can be recognized. The FAO (2004) calls these dietary convergence and dietary adaptation. Dietary convergence refers to the increasing dietary similarities in energy and nutrient composition between those of the industrialized countries and the diets of the emerging middle classes of developing countries. Both are characterized by overnutrition. Dietary adaptation reflects the change of diet under pressures from the urban lifestyle. It is characterized by a greater reliance on staple grains such as wheat and rice, as well as the increased consumption of meat, dairy products, edible oils, salt, and sugar, and a low intake of dietary fibre. It is of importance to realize that dietary convergence takes place within the indigenous food culture and the modern urban diet is not just a copy of Western food culture. Dietary acculturation is taking place and is linked with globalization and urbanization. People increasingly incorporate new foods and dishes into their diet. This process is not limited only to urban areas; similar developments can be found in rural areas.

Nutrition in transition

Why are urbanites so receptive to new food and food products when compared to rural populations, taking income as a limiting factor into account?

- A basic difference with the rural zones is that nearly all food in the city has to be bought. Urban agriculture exists and may contribute to food availability for a number of households, but most food has to be purchased in an urban environment. The urban

household is no longer a primary food-producing unit, and its members are modern consumers in the making. Urbanites can more easily make their own food choices.
- Compared to rural areas, the city is supplied with a much greater variety of food throughout the year through the marketing system. Seasonality in the food supply is less profound. The urbanite can choose from a whole range of foods.
- Long working hours and separation between residence and place of work make the urbanite receptive to food that is easy to prepare, saves time, and also takes less fuel. A related aspect is that housing conditions are often not favourable for the traditional food preparation methods at home. Pounding food in an apartment block or using traditional stoves will create tensions with neighbours. Food advertising and other marketing activities, supported by an efficient network of retail outlets, small shops, and market places stimulate the purchase of all kinds of food.

Food consumed within the urban household is, generally speaking, traditionally oriented, whilst new foods and dishes are consumed more outside the home at work, for instance, or at school. Food vendors, street foods, roadside restaurants, and the canteens in industrial establishments all serve the urban consumer. The role of street foods in urban nutrition will be further discussed in section 4.2.

The dynamics of the urban food lifestyle is strongly influenced by three reference points: the traditional rural background, urban socialization, and individualism (Bricas, 1994). Much of the food preparation methods and dishes in the household resemble the original rural and cultural background of the family. Ties to the place of origin become looser with the second and third generation living in the city. As part of the process of urban socialization, new kinds of work and practises, new social relations and daily exposure to an influx of new ideas and advertising messages force the urbanite to alter a part of his food habits. These elements are markers of the urban way of life.

A phenomenon closely associated with urbanization is the rise of individualism. The family bond is becoming weaker and individuals can escape more easily from family control. Outside the household, the urbanite can make individual food choices, not hindered by relatives. Meat has to be shared with others at home, but in a food stall or at a roadside restaurant one can just take it. This process of individualization also leads to another social phenomenon, which is the weakening of care for the elderly by younger family members.

Besides changes in food consumption, the transition of urban food habits includes changes in food consumption, changes in the place of food purchase and consumption, and the phenomenon of eating out. A gradual shift is taking place in better-off households, who once were buying food from the market and neighbourhood shop but now shop at various kinds of supermarkets. Eating out is for many households a necessity when setting off for work or during working hours. People will certainly enjoy their food, but eating out as a form of entertainment is new. Eating out for leisure is a habit that arises among the upper middle-class.

Socio-economic differences

Compared to cities of the industrialized countries, most city councils in developing countries have neither the means nor the ability to provide public housing for low-income groups and the poor. Apart from the often well-planned central business district and residential areas for the better-off, the large part of the urban population lives in the crowded squatters or spontaneous settlements; the urban slums, shantytowns, *bidonvilles* in the francophone countries, or the *pueblos jovenes* in Peru, as they may be called. Often the shantytowns are considered as no-go areas by the authorities, and even NGOs find it is not always easy to implement nutrition or health activities. The term slum is indiscriminately used, but a slum is in fact an old urban area in decay, whilst shantytowns are new, spontaneously grown settlements. Table 4.2 provides a summary of the food habits and related nutrition of the different socio-economic classes.

The higher and middle classes have a diet which can easily lead to cardio vascular diseases, certain forms of cancer, and other chronic diseases. High in energy and fat, refined foods combined with less physical activity are characteristic of affluent societies and basically do not differ from the industrialized countries, reflective of the trend toward dietary convergence. The opposite is to be found in the slums, where access to food is a major problem because of poverty. Modern urban amenities such as safe pipe born water, sewer systems, and refuse collection are insufficient or lacking. Food hygiene at the household level is a serious problem and gastro intestinal disorders leading to malnutrition are widespread.

It is striking and promising that shantytown dwellers have taken many initiatives to improve their living conditions, despite all the urban constraints. In Lima, Peru, people are involved in upgrading their make-shift houses and the environment after the city council legalized occupation of the area and acknowledge the property rights of the plot holders. Well-known in the field of food and nutrition self-help programmes are the activities of the *Commedoros Populares*, a people's communal kitchen providing low-cost meals and educational activities for the neighbourhood (UNICEF, 1994). The kitchens sprang up from organized land invasions of people looking for space to construct urban housing. The problem of insecure residency in those areas strengthened grass-root organization and women's participation.

Women play a crucial role in the management of the communal kitchens. Other promising urban nutrition activities are to be found in other cities such as Mexico City,

Table 4.2. Food habits and nutrition issues among different socio-economic groups living in urban areas in developing countries.

Socio-economic groups	Food habits and nutrition issues
High:	
Industrialists, top managers, high-ranking civil servants.	Varied diet such as meat/fish, dairy, eggs, vegetables and fruit, high consumption of imported food. Restaurants and modern fast food. High intake of energy, proteins, and fat; low physical activity, obesity. Increase of diabetes, cardiovascular diseases, other diet-related non-communicable diseases.
Middle:	
Professionals, lower civil servants, small entrepreneurs.	Much in common with the high socio-economic groups, purchasing power is a limiting factor. Good quality street foods, modern fast food. Moderate physical activity, obesity is an issue.
Low:	
Mainly unskilled irregularly employed workers; high rates of unemployment. Workers informal sector. More than 40% of total urban residents living in slums.	Food insecurity. Monotonous starchy diet. Low quality street foods. More than 60% of total expenditure on food. High rates of infant and maternal malnutrition.

Dhaka in Bangladesh or in Nairobi, Kenya (UNICEF, 1994). Compared to Lima, these programmes are less imbedded in the community.

With the assistance of the World Bank, governments in the 1960's and 1970's started to initiate programmes for site, services and settlement upgrading. The idea was that the government should provide land for self-help housing, and that infrastructure such as roads, schools, *etc.* would follow. This all was meant to be low cost, but it did not reach the urban poor. The World Bank paid new attention to the complex issue of urbanization in 2000 (WorldBank, 2000). Important from a food and nutrition point of view is the renewed attention on the scaling up of amenities of the poor. Access to safe water in a clean environment is one of the pre-conditions for healthy nutrition.

A number of key issues related to the urban poor living in the slums and squatter settlements are (IFPRI, 1998):

- Where do the urban poor get their food from and at what costs? Poor consumers are often forced to buy food in small quantities from neighbourhood stalls and small shops. Buying small quantities is rather expensive and poor consumers often have to be contented with cheap and low quality food. The purchase of street foods is often an alternative (see section 4.2).
- What are the constraints in earning an adequate income in such a way that access to food in comparison with other needs can be obtained? The informal sector is the only way for most of the poor to find employment opportunities.
- Does urban agriculture have an impact on the food situation of the household? Urban agriculture is part of a coping strategy of household food security in some cities where wasteland is still available.

Population growth in the urban areas of developing countries will be at an average of 2.3% per year for the period 2000-2030, whilst rural areas are expected to grow slowly with an average of 0.1% (UN, 2000). There are indications that the share of the urban poor in overall poverty is increasing. Without appropriate food policy measures aiming to improve access to food, malnutrition among the urban poor will remain considerable (IFPRI, 1999).

Changing life styles

The shift from coarse grains to non-traditional grains in the diet is closely associated with urbanization and is a widespread phenomenon. There has been a remarkable shift from sorghum and millet to rice in urban Burkina Faso, for example. Rice is mainly consumed from food vendors at midday when both men and women can't return home for a family midday meal for practical reasons (Kennedy and Reardon, 1994). The urban diet of the Melanesian population of Papua New Guinea has changed profoundly with the modernization of lifestyle (Hodge *et al.*, 1996). The starchy staples yam, cassava, and banana are largely replaced by imported staples such as white rice and wheat consumed in the form of bread. The continuing transition to a modern urban diet in several tropical countries is likely to lead to increasing dependence on imported food. The growing intake of refined food in many communities of Papua New Guinea, including sugar and fat combined with less physical activity, will increase the prevalence of non-insulin-dependent diabetes mellitus and cardiovascular disease.

The city is the place of changing lifestyles. Traditional food culture in the city is faced with new ideas coming from outside about what to eat and not to eat. It is a question of time, but the emerging middle-class will be influenced by globalization. An interesting case study on how a tightly constructed food culture has changed is the study by Pat Caplan (2002) on middle-class households in Madras, India, from the 1970's to the 1990's. Food in the 1970's was usually purchased from shops, including ration shops, markets, and street vendors and was rarely packaged. Some processed foods were available, including bread, biscuits, cake, and tinned food such as baked beans. Most of this processed food and soft drinks were made by Indian food companies.

During the 1980's, profound economic changes began in India, which included a more liberal economy and a loosening of the import substitution policy, a process that has accelerated since 1991. Indian food manufactures have moved increasingly into the production of elaborate packaging for their products. The middle-classes demanded high quality products and convenience food. There was a dramatic growth in self-styled supermarkets, patronized by the wealthy upper middle-classes, often young couples shopping together. Multinationals entered the food economy and the indigenous soft drink industry has been largely bought out and replaced by Coca-Cola and Pepsi-Cola. Industrially processed food is now heavily advertised all over India. Eating out as a form of entertainment became popular, increasingly in places where foreign foods are served such as Chinese, Thai, Mexican, and Italian cuisine, including pizza, although also in more traditional restaurants with Brahmin cooks.

The process of globalization will create a demand for new food and ways of accommodating the new food into the recipient food culture. On the other hand, there is a resistance to change. In the case of Madras, south Indian food and the regional Indian cooking continue to be served on important ritual occasions such as weddings and *upanayanam* (sacred thread ceremonies). Some ideologists are of the opinion that globalization will bring economic independence and self-reliance to an end. The introduction of the first beefless MacDonalds Big Mac in its first restaurant in Delhi in 1996 received strong opposition. The firm consulted Hindu religious leaders on how to put mutton on the menu instead of beef, using all Indian ingredients (Economist, 1996). International fast food chains serve affluent consumers, but are at the same time criticized as foreign intruders in the national food culture and contributing to obesity (Watson, 2006).

The local food industry in many low-income countries is often not in a position to produce sufficient high quality food for specific urban demands, food which is nutritive, high quality, easy to prepare and store, and affordable. When reaching the middle classes, it faces fierce competition from imported food and food products. Small-scale food manufactures, often micro-entrepreneurs, already play an important role, but they need more support from government policy. The provision of micro-credit and applied research in food technology is a means to stimulate the local food industry for the benefit of the low-income urban consumer (Lopez and Munchnik, 1997; Trèche *et al.*, 2002).

4.2 Street food

Street foods and the city

People in the city tend to prepare less food at the household level and consume increasingly ready-prepared food from street food vendors, roadside restaurants, canteens, and other food selling places. Among all the possible food outlets, street foods are the most dominant form in cities of developing countries. What are street foods? Street foods can be defined as ready-to-eat foods and beverages sold by vendors and hawkers, especially in streets and other similar places (FAO, 1990b).

The significance of street foods to the urban way of life can be approached from three levels:
- As an employment opportunity, particularly for women in the informal sector, permitting the development of small-scale urban entrepreneurship. Children are also employed in the street food sector by selling bread, sweets, or cleaning the utensils and stools of makeshift roadside restaurants. The employment of children remains a point of concern.
- Food technology and the necessity to give attention to food control, hygiene, and other aspects of quality. Street foods may stimulate the development of an artisanal or small-scale food industry and appropriate food technology.
- The nutritional dimensions of street foods, such as their contributions to nutrition and health of the consumer, and the extent to which street foods correspond with consumer needs.

Street foods are usually consumed as a snack or meal on the roadside, but also at home. People may buy a ready-prepared staple and take it home by adding a self-prepared accompaniment such as a stew or other side dishes. Street food can also be used as an ingredient by adding at home ready prepared fish or meat to the staple.

Although women prepare food at the household level, the situation at the professional level is different. In countries where women play a clear role in public life, most street food vendors are women. The participation of women in the street food sector varies from 94% in Nigeria to 24% in Egypt and 1% in Bangladesh (Chauliac and Gerbouin - Rerolle, 1994). There is also a gender aspect to the kind of street foods being sold. For example, the sale of meat, bread, and coffee with milk (*café au lait*) are men's business in the Francophone countries of West Africa, while cereal porridges and fried foods belong mainly to the domain of women. Roasting meat and the sale of roasted meat is a man's job, for instance, as it is the case in the Maghreb countries as well. However boiling meat as part of a dish is the task of women (Maarmar, 2003).

The street food sector is characterized by a hierarchy of food selling. It ranges from the ambulant food vendors with simple products such bread, sweets, or doughnuts to semi-permanent and even fixed stalls, kiosks, or makeshift restaurants selling whole meals and dishes. Well known in Francophone West Africa is the *maquis*, a fenced courtyard with tables and chairs, serving sophisticated African dishes for better-off clients.

Food hygiene and quality is a point of concern in the overcrowded streets and quarters. Apart from microbiological problems related to inadequate food preparation methods and a lack of safe water, various other toxicological issues exist. Dirt from the roads and car exhaust fumes spread freely over the kitchen and the dishes. A study in Nairobi showed that when food vendors had a basic knowledge of food hygiene, they could not bring this knowledge into practise as basic amenities such as safe water and refuse collection were lacking in slum areas (Mwangi, 2002). It is not uncommon that urban authorities have a bias towards street foods. Street food vendors produce waste, congest the streets, and sell contaminated foods. Some authorities consider street foods as part of underdevelopment and that the food sellers should be removed from the central business district of the city. Most vending sites are self-allocated and not furnished with amenities. As vending sites are often tolerated but not legalized by urban authorities, vendors are not willing to risk their capital by investing in a venture that the city authorities might demolish at any time.

Consumer and nutritional dimensions

The variety of street food products sold and consumed is enormous. Typical products can be found in every country, ranging from simple snacks to elaborate dishes, including cereals, fruits, vegetables, meat, fish, biscuits, and all sorts of beverages. Street foods are available at places where they are required, *e.g.* around factories and offices, schools and colleges, market places, transit areas, and also in the crowded residential areas. Consumers are of all ages and different socio-economic classes. The nutritional value varies with the type of food and depends on the quality of the ingredients used. The nutritional value of street foods needs to be related to the total food intake. In Nigeria, street foods are found to provide up to 59% of the daily energy and nutrient intake of urban market women (Oguntona and Tella, 1999). The contribution of street foods and other non-home prepared foods to the energy intake of consumers ranged from a median of 13% to 36% for subjects living in low-income areas of Nairobi. These figures are the same for slum dwellers living in Hyderabad, India, and Haitian school children (Van 't Riet *et al.*, 2001).

The place of street foods in the diet varies considerably. In many countries it provides breakfast and the midday meal, while the evening meal is often consumed at home. As a source of a snack or in-between meal, it has the same function worldwide. In the poor households of Nairobi, street food is mainly consumed as breakfast, snacks, and to a lesser extent, as a midday meal (Van 't Riet *et al.*, 2001). Most workers in the Nairobi industrial area get their lunches from street food vendors (Mwangi, 2002). See also Table 4.3.

Street food is an indispensable food for the urban consumer, not only for the urban poor, but also for the middle-income groups. It is time and fuel saving. Low-income households have insufficient cash to buy all the raw ingredients and fuel to prepare a meal at home. In this way they are forced to buy already prepared food on a day-to-day basis.

Industrial workers with fixed income in Nairobi could afford to buy a varied lunch, including meat, fish, and vegetables from female food sellers (Mwangi, 2002). It is logical that the majority of urban workers cannot depend on food prepared at home, as they must contend with long working hours, workplaces that are far from home, and a lack of rapid transport, as well as the increasing costs of transport (Sujatha *et al.*, 1997). Trends in street food vending in Nairobi suggest that street foods were originally convenient foods for low-income groups and the poor. This indicates that street foods are not necessarily convenient due to women's opportunity cost of time, but mostly due to their affordability and the meagre earnings of the poor (Mwangi *et al.*, 2001). Generally speaking, poor households are not in a position to prepare the same meal at lesser costs than the food vendor. It is thus more cost effective for these households to buy from food vendors.

Table 4.3. Proportion of consumers buying street foods for various meals and snacks for at least four days in a week in three study locations in Nairobi, Kenya.

Type of meal/snack	Korogocho Slums (n=84) %	Dandora Low-income area (n=104) %	Industrial Area (n=54) %	Total (n=240) %
Breakfast	76	68	33	63
Major morning snack	20	27	4	19
Lunch	50	56	95	63
Major afternoon snack	38	39	15	33
Dinner	31	35	4	27
Petty snacks	58	62	19	51

Source: Mwangi, 2002.

4.3 Urban agriculture

Intensive urban-rural interactions are a characteristic phenomenon for most cities, having far-reaching impacts on food and nutrition among both urban and rural populations. The interaction takes place at two levels: the urban-rural links (sections 5.1 and 5.2), and the urban agriculture within and around the city. In urban and peri-urban areas, relatively sizable tracts of land are used for food production. Urban agriculture provides an important contribution to the urban food supply and also provides a source of income. It involves food production in open spaces not yet built up and can be found between town squatters, on idle public land and land considered unsuitable for building, and along roads, railways, and airports. Small home gardens also play a role in food production. Urban agriculture is not exclusively an economic activity of poor and low-income households. Middle-class households may invest in food production in the peri-urban areas (Smit *et al.*, 1996).

Urban agriculture is important to the food supply and livelihood of many cities. It has been estimated that 15-20% of the world's food was produced in urban areas, but this percentage is on the increase. On a worldwide basis, 800 million people are engaged in urban agriculture. Of these, 200 million are considered to be producing for the market, employing 150 million people full time. The percentage of urban families involved in urban agriculture may range from 16% in Cairo, Egypt to 45% in Lusaka, Zambia (Anonymous, 2002). The major nutritional dimensions of urban agriculture are:

- From a nutritional point of view, the energy intake of food produced by urban agriculture is limited in contrast to its contribution to the qualitative aspects of food. Major products are vegetables and fruits (micronutrients) and some small-scale animal husbandry of poultry, ducks, pigs, and fishponds (proteins). A striking example is Hanoi, Vietnam, where 80% of fresh vegetables, 50% of pork, poultry, and fresh water fish, as well 40% of eggs, originates from urban and peri-urban areas. In Accra, Ghana, 90% of the city's fresh vegetable consumption is from production within the city (Anonymous, 2002).
- In times of civil unrest and economic crisis, urban agriculture is of direct importance for coping with food shortages. This has been the case in a number of African cities, where the number of people engaged in urban food production has increased rapidly (Foeken *et al.*, 2004).
- Urban agriculture is faced with a number of constraints threatening nutrition and health. In the first place, there is the issue of uncontrolled application of pesticides. Polluted water will create environmental risks. Faeces-containing manure is used in many cities, which may cause contagious diseases. Animals kept in crowded pigsties and henhouses close to human settlements may cause human health risks. For example, the first documented infection of humans with the avian influenza virus occurred in Hong Kong in 1997. New outbreaks of avian influenza in East and

Southeast Asia caused several deaths in 2003 and 2004 (WHO, 2004). Humanity has become vulnerable to cross-species illnesses, thanks to increasing population density, intensified livestock production, and the mass transport of people, food, animals, and animal products. Besides influenza, other diseases such as a bacterium typical for pigs, the *Streptococcus suis*, is able to cross the species barrier and to infect people. Diseases arising from contact with wild animals (handling of wild meat), such as the Ebola virus in tropical Africa, will reach the cities. SARS (Severe Acute Respiratory Syndrome) also arose from contacts with wild animals. The illness first appeared in rural areas of China in 2002 and reached the city of Hong Kong in 2003, spreading later to other continents. (Garrett, 2005; Karesh and Cook, 2005).

Urban agriculture will create employment opportunities for women in the city. However, a major point of concern continues to be the uncertainty and lack of legal security over land rights, and the right to continue on the same plot with food production. In the forever-expanding cities, urban farmers face the problem of being evicted from their land. Just as with street foods, urban authorities should take notice of the importance of urban agriculture in urban development planning and projects.

5. Rural areas on the move

5.1 Urban-rural food links

Urban and rural areas are closely linked with each other by migration to the city and by the spread of urban influence into the rural areas. From early times on, cities have endeavoured to bring rural areas under their political and economical control to ensure a steady supply of food to the city. Political and economic power in most countries is centred in the city.

Rural-urban migration

Rural-urban migration presents the danger of young able-bodied men leaving the village, reducing the food production labour force and leaving women, children, and the elderly behind. A relatively new phenomenon is that young women are also leaving to get a job in the city or to find employment overseas. The Philippines is an example of a sizable migration of both men and women from the rural areas to the city and to the Arab oil states. Though it is often stated that traditional farming has a surplus of labour, this is not true with the peak demand at certain periods of the year, *e.g.* during weeding and harvesting when much labour is required. With an increased workload placed on rural women when men are absent, nutritional care, education of children, and household cohesion will suffer. Even when men show their sense of responsibility by sending sufficient money back home to their households, their absence cannot be fully compensated. These phenomena have been observed in many parts of the world. The impact of the absence of younger men and many of the younger women on food farming was already observed by Polly Hill in Ghana in the 1970's (Hill, 1978). The efficiency of farming was reduced because most remaining farmers were middle-aged or elderly and the supply of day-labourers was inadequate. Husbands and wives have separate but complementary farming tasks. For instance, men always clear the forest and bushes and prepare the fields for cultivation, whereas women weed and do the food marketing. Spouseless women find it difficult to get someone to do agricultural work considered appropriate for men.

A large-scale study on migration and remittances carried out by the European Commission of the EU showed that the remittances of migrants working in the European Union can be quite considerable. On average, households receiving remittances in Egypt and Turkey received $400 per year. In Morocco, it even amounted up to $1,350. Most of the money is used for food, clothing, and housing. The amount of money sent depends on the composition of the household. Female-headed households with young children get more than households headed by men, particularly men above 65 years of age. The study concludes that migrants are more inclined to send remittances home to young families than to feed the elderly (Fokkema and Groenewold, 2003). The traditional family structure in rural Yogjakarta, Indonesia, including care for the elderly, is still rather stable despite migration of young people to the city or other parts of the country.

Traditionally, Javanese children, as in many other societies, are responsible for taking care of their older parents. Nowadays families are smaller and young members migrate, relatives live at larger distance from each other and women are increasingly involved in off-farm employment. As a result, traditional support mechanisms can no longer be relied upon for the care of older people. Older Javanese prefer to live with their adult children in extended families. There are indications, however, that certain groups of vulnerable people can be found. If women do receive remittances, the amount will be lower than those received by men. The households with the oldest members receive on average the lowest amount of money (Keasberry, 2002).

5.2 Cash cropping and food farming

Pure subsistence farming, or the production of food and other products only for household consumption, can only be found in very remote areas. Farming in most rural areas of developing countries is characterized by the production of food crops for household use and crops (food and non-food) to sell. Cash cropping can nowadays be found in remote areas. Money in the rural areas has become a necessity for coping with daily life. Cash is needed for both food and non-food expenses, such as:

- The purchase of food products such as salt, tea, coffee, sugar, meat, or fish, all depending on the local conditions and food culture. Related to food are the expenditures on matches and sometimes, even in rural areas, on fuel for cooking (firewood, charcoal).
- Some money has to be set aside for buying food in the lean period of the year, the time before the new harvest.
- The purchase of clothing, soap and other related articles.
- To meet social obligations such as hospitality, dowry, weddings, and funerals.
- To pay for children's school fees (school uniform, pens, school note books) and health services, even when these provisions are meant to be free for everyone.
- To pay taxes for the local or central government.

The decision-making process in rural households includes the amounts to be spent on food, including some money set a side for the lean season, and other non-food items that are needed. With economic development moving towards cash crops and marketing, there is a kind of competition going on between what should be spent on food and what should be spent on other needs. By devoting more time to cash crops, there is a potential danger of replacing labour-intensive food crops with food crops which require less labour, but are nutritionally inferior. Generally speaking, land is scarce, so cash crops may occupy land which is needed for household food security. The issue is often how the cultivation of tea, coffee, sugar and cotton can be combined with food crops. Will the cash income be sufficient to compensate for the lesser availability of home produced food? There may be a temptation to spend too much time and effort on non-food items.

Low world market prices, sharp market fluctuations, and limited access to markets of the industrialized countries all prevent a rise in standards of living and better nutrition for rural populations. An extensive study carried out in the 1990's concluded that the commercialization of agriculture had not resulted in dramatically negative effects on food and nutrition among rural populations (DeWalt, 1993). However, the issue of cash crops versus food crops needs continued surveillance. West African cotton farmers have to compete on the world market with highly subsidized low price cotton from the United States. As a result, farmer cash income is very low, inadequate for the improvement of living conditions and nutrition. In addition to non-food crops, farmers produce staple crop, vegetables, fruit, and animal food to the urban markets. In Africa south of the Sahara, there has been a long, ongoing process where the age-old traditional food crops, yam, sorghum, and millet are being replaced by cassava, maize, or rice (see Figure 5.1).

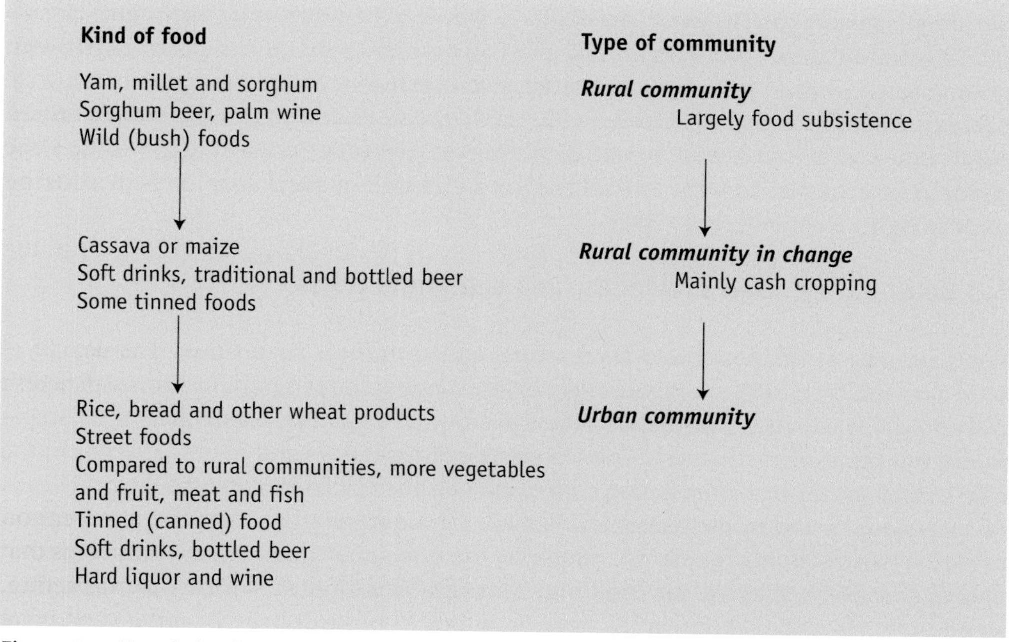

Figure 5.1. Trends in the shift of major food stuffs in Africa south of the Sahara.

There is a tendency in Mexico and Central America to replace maize with rice and wheat consumed in the form of bread. Along with rising living standards comes the tendency to replace traditional unrefined and coarse staple foods with refined ones for reasons of taste, convenience, and also prestige, but to the detriment of nutritional quality.

A category of rural populations that need much attention are landless farmers, farmers who have no land at all, even in many cases building their houses on rented plots. In the densely populated countries of Asia such as India, Bangladesh, Sri Lanka, and on the island of Java, Indonesia, this amounts to more than 30% of the rural population. Own food production gives no, or a very minimal contribution in maintaining household food security. As casual labourers, these households are living in a state of continuous food insecurity. In rice farming communities of the Philippines, it is common practise for landless households to provide weeding services when rice plots are ready for weeding. Under the *hunusan* arrangement, landowners give landless farmers the exclusive rights to harvest on the same plot the weed and to receive a share from the rice harvest. The rice share is an important buffer for many landless households during the lean period of the year. Or, as a female worker said: "As long as my husband and I do not fall ill and there are no natural calamities, my family will not go hungry because our share from the harvest will be more than enough for us" (Balatibat, 2004).

Who is better off nutritionally, the rural or the urban poor living in the shantytowns? No simple answer can be given. An earlier review on the nutritional status and dietary intake from different countries in Asia and Latin America found comparisons between poor urban and rural populations inconclusive (Atkinson, 1992). A recent comparison between Angola, Central African Republic, and Senegal demonstrates that simple urban/rural comparisons mask wide spread disparities in subgroups according to wealth. Poor children in urban areas were as likely as poor children in rural areas to be stunted or underweight (Kennedy *et al.*, 2006).

5.3 Seasonality, food shortages, and economic crisis

Food security at the household level means access to food at all times. The degree of food availability in rural areas is strongly determined by the geographical zone where the community is situated. Seasonality is less profound in the city, particularly with staple foods, but prices may fluctuate. The seasonality of vegetables and fruit is also felt in the city. Urban authorities often have a close watch on the prices of major foods for reasons of social policy and to prevent social unrest. Of a different nature is the lack of food caused by an economic crisis. For example, the economic crisis in Asia in the second half of the 1990's affected the food and nutrition situation of low-income urbanites. In all these situations, households develop a food coping strategy in order to survive (Maxwell, 1996).

Seasonality

Climatic seasonality and the ensuing agricultural periodicity affect many aspects of daily life. Some of these effects are changes in labour needs, food consumption, nutritional status, health, and vital events as well as various aspects of social life. Others have to be regarded as being more in line with coping mechanisms, notably in the economic sphere (trade, prices), the demographic sphere (migration) and the political sphere, such as government food interventions (Foeken and Den Hartog, 1990).

Seasons in the temperate regions of the world are primarily determined by fluctuations in temperature and sunlight. Precipitation determines the nature of the seasons in the temperate regions of the world. On the basis of rainfall distribution throughout the year, three types of tropical climates can be distinguished.

- Climates with no real dry season are to be found on and near the equator.
- Climates with two rainy and two dry seasons, the so-called bimodal climates, are found close to the equator.
- Climates with one wet and one dry season, the unimodal climates, are to be found further from the equator.

From a food and nutrition point of view, the distinction between bimodal and unimodal climates is an important one. In general, people living in a bimodal climate experience less seasonal stress than those living in a region with only one rainy season. In the savannah zones of tropical Africa, the effect of the long dry season on food availability is sharp. At the end of the dry season and the beginning of the rainy season, when the household grain silos are nearly empty, the period is called in many of the local languages the *hungry season* or, in French, as *période de soudure*. Seasonal food shortages can be of such a magnitude that seasonal body-weight losses of 2-6% of the total body-weight occur (Durnin *et al.*, 1990; Van Liere *et al.*, 1994). The failure of the rain to come on time in monsoonal Asia can cause serious food shortages, sometimes leading to famine.

Seasonality and coping behaviour

Households living in regions with seasonal food shortages have developed a coping behaviour over the course of time. Coping behaviour can be defined as a strategy to resist a problematic situation, in this case increasing food shortages. Such households are more prepared and know much better what to do than households living in regions where food shortages seldom occur. Households in a region with marked seasonality take two kinds of measures: preventive measures and the actual coping with shortages when the problem has arrived. Preventive measures might be, for instance, planting more drought resistant crops and making the household grain silos and other food stores as full as possible after harvest. Old rural wisdom holds that after harvest, the silos should be pilled up in such a way as to cover the expected household needs until the new food crops can be harvested. Only the surplus can be sold for cash. However, the pressing need for cash may cause too much of the household food stock to be sold. When running out of stock, additional food has to be bought at high prices.

Besides an adjustment of the diet, an adjustment of a physiological nature takes place, which is the reduction of physical work. People may diminish activities by cutting out specific tasks, for example by reducing time allocated to repair works or by slowing down the pace, or diminishing the intensity of the activity. In contrast to the dry season, when people do not have to work in the fields, the situation becomes problematic during the rainy season when the fields have to be prepared and food stocks are diminishing. A temporarily undernourished population will prepare the fields for the coming new agricultural season at the end of the dry season and at the beginning of the rainy season (De Garine and Koppert, 1988; Sanaka Arachi, 1998).

Coping with food shortages

The gravity or degree of food shortages at the household level is determined by the remaining available quantity of food and the expected duration of the shortage. This may happen during what seems to be an usual seasonal shortage, but which slowly develops into a chronic shortage; a famine. It is not easy to assess at the household level whether or not the seasonal shortage is leading to famine. At the policy level, the Global Information and Early Warning Service of the UN and national early warning systems have done much work in predicting food shortages so that appropriate measures can be taken. Households confronted with food shortage will make a cognitive appraisal of the encounter. The appraisal is based on two questions: what is at stake in this specific encounter, and what can be done, or what the possible options for coping with it are. A hierarchy of coping strategies can be constructed from the available literature. This hierarchy is mostly determined by the nature and duration of the shortage (see Table 5.1). In reality, however the hierarchy of measures are not strictly separated and regional differences occur (Den Hartog and Brouwer, 1990).

In situations where food shortages occur with a distinct regularity, the first coping strategies will mainly be focussed on the diet. These imply a decrease in food consumption by reducing the number of meals per day, reducing the portion, and diluting the meal with extra water, such as a porridge or stew to suppress feelings of hunger. When a food shortage lasts longer, non-conventional foods will be consumed. These foods, mostly of vegetable origin, are collected in the fields and are known as *hungry foods*. These foods are omitted from the diet when common food is abundant again. It is not unusual for groups of women in the savannah regions of tropical Africa to search for the seeds of wild grasses, tubers, and wild fruits, which would otherwise not be eaten. In the past,

Table 5.1. Food shortages and a hierarchy of coping strategies by households in rural areas.

Type of coping	*Specific action*
I Seasonal shortages	
Reduction of quantity	Measures • Reduction in number of meals • Reduction of portions • Diluting meal with extra water • Adding inedible substances to the meal
Adjustment of dietary habits	Consumption of unconventional foods • Famine foods, *e.g.* plants and animals not eaten otherwise • Consumption of sowing seeds
Using up cash	Purchase of food (at high prices)
II Shortages of a chronic nature	
Selling of property	Selling jewellery, clothing (gender issue) Selling cattle, land (impoverishment)
Roaming for food	Lending money for food (high interests) Borrowing food from other households Wandering in search for food in other areas Raids
Migration	Temporary migration to other areas Boarding out of children elsewere
Religious measures	Prayer and magic (*e.g.* rainmaking)

cassava was introduced under colonial administration in Zambia and on Java, Indonesia, as a kind of reserve crop to be used in periods of shortages. When shortages further continue, people may turn to the extremely hazardous measure of consuming seeds put aside for sowing. Doing so could mean they are headed towards a new disaster. However, households try to limit the consumption of seeds for sowing as much as possible. The whole situation can be further characterized by a loss of body weight and a deterioration in health status. The first signs of disintegration can also be noticed at the social level; rules of hospitality and reciprocity between household and individuals become looser and forms of aggressiveness develop.

Long during food shortages

A seasonal food shortage may evolve into a chronic situation from time to time, eventually leading to famine. Adjustment of dietary habits will cease to give any relief. Households have to turn to other strategies, such as selling property and borrowing money or food, often at very high prices, and selling cattle or women's jewellery. The question remains, however; who will first sell these properties, the husband or the wife? In the savannah regions of Africa it is not uncommon to borrow a bag of maize from a merchant, which has to be paid back in kind with two or even three bags of maize. Households thus pay a high interest in kind.

Poor households will start roaming for food, to be followed later by the better-off households. Household members will start wandering in search for food. This search can end up in predatory-like expeditions affecting less badly hit regions. To meet a continuously worsening situation, people may turn toward religious support in the form of prayers to God or begging the gods for help. Others may try to take the course of events into their own hands by turning to magic. The so-called rainmakers of East Africa fall under this category of measures. Although religious measures are mentioned last in table 5.1, people will often turn to prayer in an early stage of the dry season.

Gender and food shortages

Women in most societies fulfil a central role in supplying, preparing and distributing food among the different household members. Women carry direct responsibility for the nutritional care of young children. The role of men in rural areas in nutritional care is limited and mainly concerned with food production. Seasonal food shortages imply a yearly recurring additional care and burden for women. Even under difficult circumstances, women are expected to be able to prepare a meal. How will intra-household food distribution be arranged under distressful circumstances? Seasonal food shortages mostly coincide with a lack of sufficient drinking water for both man and animals. Water is an indispensable basic need of the human being. Fetching water is a task of women and older children. When food availability decreases due to drought, such as in the savannah zones, water needs to be fetched from longer distances.

There is a tendency in many societies to favour the male members of the household during food distribution. In periods of relative abundance, there is usually enough food for all members regardless of how the food is distributed. When the household food stocks are diminishing, will food distribution be at the expense of women? Will women go to great lengths by saving their own food to give their children and husband? These questions are not easily answered. No firm conclusion could be reached that men are favoured with food allocation over women and children in the different cultures of

Malawi and Bangladesh (Wheeler and Abdullah, 1988). The rural people of Malawi and Bangladesh appear to have an accurate perception of adult's and older children's needs. Some discrimination against young girls does occur in Bangladesh, however. In times of food shortages, parents are concerned about giving their children as much as they possibly can. A study on food and nutrition security in the Philippines indicated that, when there is enough money, the food preferences of the husband and the children were usually taken into account. When there is a food shortage, the mother tends to distribute the food equally among other members of the household, except for the father, who is given a larger portion of viands, if they are still available (Balatibat, 2004).

The workload of women during food shortages is determined, among other factors, by the following (Jiggins, 1986; Balatibat, 2004):

- The seriousness and duration of the food shortage.
- The assistance women can get from their children or other women in the household; the willingness of the husband and other men to suspend the gender-specific role patterns in food and nutrition or, simply, to perform women's tasks in stressful situations. In poorer households, men will have to assist sooner in women's tasks than in relatively richer households.
- Power relations in decision making about how household resources should be allocated for food purchases will further determine the struggle for maintaining household food security; who owns the household resources such as land, cattle, jewellery, and clothes, and who is authorized to sell possessions for food when needed. Jewellery and clothes are often a savings in kind among women. Forced sales of these goods may deteriorate the position of women.
- The physical condition of women, which will be weakened towards the end of the dry season and further deteriorate during the rainy season when much agricultural work has to be performed.

5.4 Fuel scarcity, food preparation and nutrition

Most foods have to be cooked in order to make the food digestible. One of the major innovations of early man was the application of fire to food preparation. The full control over fire by Homo sapiens was a major step in human advancement and human nutrition (Abraham, 1986). Fire (heat) for food preparation made a large range of plants and animals digestible for man by breaking down food cellulose and improving bio-availability of nutrients. Cereals and other seeds can only be consumed in quantity by means of heating. The stages of food preparation development are: first roasting on an open fire, followed by boiling food in water in vessels, and later the technique of baking in an oven. In some civilizations of the Pacific islands, traditionally large meals and feasts are prepared in a temporary kind of stone oven. A fire is built on a flat layer of stones, when the fire has burnt out the hot top stones are removed and the flat layer swept clean.

Food wrapped in leaves will be placed on the layer and covered with hot stones and some more layers. The food is boiled by dropping water on the hot-red stones, which are lifted with bamboo tongs.

The main fuel used for cooking at the household level in rural areas of developing countries is usually wood, supplemented with crop residues or animal dung, in particular cow dung, such as in the Indian-subcontinent. Wood is now mainly collected outside the forests; from village or household wood lots, from trees and shrubs scattered over agricultural fields and wasteland. Wood collection in most societies is a woman's task, often assisted by their children. The effort of women to collect sufficient firewood for cooking, expressed in time, varies from as little as half an hour to as much as five hours per day (Cecelski, 1987). The time spent on wood collection in rural Malawi during the rainy season is 1.47 hour per day, and 2.05 hours in the post-harvest season. In some cases it can amount to 2.38 hours. During the rainy season, the prevalence of illnesses such as malaria and colds is problematic and people visit hospitals more frequently. Fire wood collection is often combined with other activities (Brouwer, 1994).

Both rural and urban households have to cope with diminishing fuel sources, in particular firewood. Firewood in the rural areas is still free, but women have to cover greater distances to acquire it, which implies an increase in workload. City households are faced with rising fuel prices. The supply of fuel wood is rapidly decreasing because of deforestation, as forests are unscrupulous cleared for local use and the export of tropical timber, farming, and overgrazing of land by livestock. The high demand for fuel wood results in the disappearance of shrubs and solitary trees. In the cities, substitutes exist such as electricity, kerosene, and coal, but for the poor urban households it is very expensive. Buying already prepared food from street food vendors is an option for saving on fuel costs. People in rural areas are encouraged to plant fast growing trees near their homes or along roadsides. Governmental institutions and NGOs promoting durable forestry are faced with the short-term interests of the timber trade and farmers looking for land (Sayer and Campbell, 2003). In many countries, projects have been initiated to promote energy saving cooking techniques and the application of alternative energy.

What will be the impact of fuel shortages on food and nutrition? The use of inferior fuel for cooking such as crop residues, animal dung, or small wet twigs collected with great effort will produce a great deal of smoke and cause respiratory and even eye diseases for those who reside near the fire, such as women and children. There are indications that food requiring lengthy preparation times will be left out when fuel costs are too high. The preparation of fewer meals and the consumption of cold or warmed-up leftovers is a frequent strategy used to cope with fuel shortages (Cecelski, 1987). Table 5.2 shows the responses of women in Malawi to a survey asking how they would adapt and change the standard diet in response to a hypothetical fuelwood scarcity.

Table 5.2 Sequence of adaptation of daily dietary patterns to a hypothetical fuel wood scarcity by women in Nicheu District, Malawi (n=20).

Dietary pattern	Adequate fuel wood availability	Marginal fuel wood availability	Fuel wood shortage	Severe fuel wood shortage
Breakfast	Porridge (phala)			
Lunch	Nsima[a] with pumpkin leaves	Nsima with pumpkin leaves		
Snack	Boiled maize kernels			
Dinner	Nsima with beans	Nsima with pumpkin leaves	Nisima with pumpkin leaves	Porridge

[a] Thick maize flour porridge, the relish contains occasionally fish, meat is seldom consumed.
Source: (Brouwer *et al.*, 1996).

Women would first replace the bean relish with other types of relish and forego less important dishes such as boiled maize kernels and porridge for breakfast. As an alternative to beans, women would cook pumpkin leaves in large amounts for lunch so leftovers could be used for dinner, or they would choose another relish which needed less fuel. The nature of such a fuelwood scarcity would lead to maintaining two meals; *nsima* (thick maize flour porridge), with a relish of vegetables for both lunch and dinner. In the last stage, only one meal was left. Women were asked at the end of the interview whether the adaptations mentioned reflected a realistic situation. Most of the women stated that they were sometimes already forced to drop or to replace dishes because of fuel shortage. However, they would always try to find wood or other fuel to prepare lunch and at least dinner (Brouwer *et al.*, 1996).

6. The food system: From production to consumption

6.1 Household food security

Food security is an essential condition that must be achieved for good nutrition and to maintain health. However, as depicted by the UNICEF conceptual framework (see Chapter 1), it is not the only condition for good nutrition, as adequate care and health combined with a healthy environment should also be fulfilled.

The definition of food security has changed since the 1974 World Food Conference from being primarily equated with dietary energy sufficiency related to adequate food production, to mere a problem of access to food. There is now a general agreement among development organizations that food security exists when all people, at all times, have physical, social, and economic access to sufficient, safe, and nutritious food to meet their dietary needs and food preferences for an active and healthy life. Adequacy is an important concept in this definition and should be considered in both quantitative terms as well as qualitative terms. An adequate diet should provide nutrients in quantities sufficient to maintain good health (depending on many factors including age, sex, level of activity, and physiological status). An adequate diet should also be safe and free from contaminants, parasites, and toxins which may threaten health. It should also be culturally acceptable and, in addition, should be palatable and be capable of providing pleasure to the consumer.

The definition of food security comprises three important dimensions. Availability of food and access to food are two essential determinants of food security. Food enters the household in different ways. A household may produce food and has direct access to food. The ability of farmers to produce food in adequate amounts and sufficient variety depends to a large extent on their access to resources like sufficient and fertile land, labour, tools, seeds, credit, agricultural services, and knowledge to grow crops and raise animals to sustain survival of the household on a continuous basis. Many rural communities also gather wild food or exchange food for labour with family or neighbours, for example, and in such a way contribute to the household food supply. Food is also purchased. Most households purchase a part of their food which they do not produce or do not produce in sufficient quantities. This represents the dimension of economic access. Access depends on household income and prices, and is therefore prone to risk, especially if jobs are lost, incomes fall, food prices rise, harvests in the rural areas fail, or relatives move to urban areas. The third dimension is related to the utilization of food. Moving from household food security to individual food security requires consideration of two factors. First, the way food is allocated within the household, or the intra-household food distribution, determines whether individuals receive enough food to cover their needs. Food may be distributed according to nutritional requirements, but sex or status determine the amount and type of food a person receives in most developing countries (see also section 6.3). Secondly, biological utilization needs to be considered, referring to the ability of the human body to take food and to convert it into nutrients that can be used by the body to maintain health.

By implication of the above, those being food insecure have lost, or are at risk of losing, the availability of and access to food, or the ability to utilize it. Groups most vulnerable to food insecurity share common socio-economic, agro-ecological, demographic, and educational characteristics related to distribution and control over resources or access to employment. Subsistence farmers who produce marginally, landless wage earners, female-headed households, households with a large number of dependants, households situated on marginal lands are examples of vulnerable groups, among which children under five and women of child-bearing age are most at risk (Ahmed Ali, 2005). However, it is important to note that vulnerability to food insecurity is location-specific and should be assessed independently for each community or district.

The concept of food security also has spatial and temporal dimensions. The spatial dimension refers to the degree of aggregation at which food security is being considered at the global, continental, national, sub-national, village, household, or individual level. It is important to note that, for example, national food security does not guarantee that all households are food secure, as this does not take food distribution within the country into account. Likewise, household food security does not mean that all individuals in the household are food secure, as some factors such as intra-household food distribution, for example, are not taken into account.

The temporal dimension refers to the time frame over which food security is being considered. Households may suffer from transitory food insecurity when the ability to meet foods is of a temporary nature. This could be cyclical (where there is a regular pattern of recurrence of food insecurity, such as the "lean" season just before harvest), or temporary, which might be the result of a short-term, exogenous shock such as droughts or floods, but also wars. On the other hand, households are chronically food insecure when they persistently lack the ability to meet food needs on an ongoing basis. This is related to the household's ability to cope with or minimize the extent and duration of food deficits and whether they are able to bounce back or to regain quickly an adequate food supply. Recurrent exposure to temporary stress with no opportunity to regain food supply may lead to a downward spiral to food insecurity of a chronic character when households deplete their resources.

There are important differences in household food security issues in rural and urban contexts (Von Braun *et al.*, 1992). Household food security in urban areas is primarily a function of the wage (relative to food prices) and of the level of employment. The miserable health environment in poor urban areas also sometimes makes the urban food security situation qualitatively different from the rural situation. Differences in calorie consumption and requirements exist between rural and urban areas, with lower consumption in urban areas partly because of differences in activity levels. Although the prevalence of food insecurity is lower in urban areas than in rural areas, urban poverty with chronic food insecurity will become an increasingly important problem in the future with higher rates of urbanization.

6.2 Food preparation and storage

The function of food preparation was at first to make nutrients biologically available for the body. Raw foods such as tubers, grains, and legumes, for instance, contain carbohydrates, which are not very readily digested and may contain toxins. After cooking, carbohydrates are more easily digested and some toxins are inactivated. Secondly, foods are prepared or processed to improve their palatability and to extend their lifetime and prevent microbial quality deterioration.

There are several ways of processing foods. On a household level, boiling, baking, and frying are almost daily practices. Insufficient fuel may contribute to food insecurity, as has been stated in Chapter 5. On the one hand, heating may make nutrients more readily available; on the other hand, losses of especially water soluble vitamins may occur. Most discussed is the loss of vitamin C and the leak of B vitamins and some minerals that occurs with boiling. Losses increase with the amount of water added and also when the food is overcooked. Another example is loss of the amino acid lysine in baking bread, due to the browning reactions between proteins and carbohydrates that contribute to the brown color and aroma of baked products.

The heating of foods may also lead to the formation of toxic components under certain conditions. Examples are the potential carcinogenic compounds in wood smoke transferred to charred meat and other products exposed to smoke. Potentially carcinogenic is also acryl amide found in fried potatoes and crisp bread. The cook is in general advised to be careful with very high temperatures, but this is easier in a modern kitchen than on an open fire outside the house.

Not only heating, but also cutting and mashing may make some nutrients more readily available. An example are the carotenoids (pro-vitamin A), which are better available from mashed tomatoes (juiced or pureed) than from whole fresh tomatoes. In contrast with mashing, losses of vitamin C, for instance, may occur. The size of losses depends on the technique applied. Training in food preparation (and hygiene) may prevent food and nutrient losses.

An example of a no- or low-heating method of food preparation is fermentation. The latter is defined as the action of microorganisms or enzymes on raw food components to cause biochemical changes. These changes may affect the nutritional quality, digestibility, safety, or flavour of food. Many products are also produced by fermentation on the household level. Examples are tofu, tempeh, yoghurt, cheese, bread, and beer. Lactic acid bacteria, which convert carbohydrates to lactic acid, thereby lowering the pH, are important in many fermented foods, including dairy products, meat, and cereal products. Fermentation with lactic acid bacteria prolongs shelf life and, especially with the application in milk, was in the past century an improvement in the safety of infant nutrition in milk drinking cultures.

Losses of food occur throughout the food chain. Some of the losses are inevitable consequence of the conversion of the primary product into edible food. However, losses also occur as a consequence of storage, some of which are avoidable. The latter include losses due to spoilage by rodents or insect pests, or due to fungal or bacterial damage to the stored crop. Fungi may produce metabolites that are toxic. Epidemics due to the consumption of rye that had been stored in damp conditions were common in the past. The grain was contaminated by the fungus known as ergot, and ergot poisoning was manifested either as gangrenous or as convulsive ergotism. Concern over the presence of aflatoxins produced by the fungus Aspergillus flavus has developed since the 1960's, at which time it was found that these metabolites were toxic for animals. Aflatoxins occur in mouldy grain, soybeans, or nuts and are toxic at very low levels of intake. The problem is that aflatoxins and many other mycotoxins are quite stable during normal food preparation or food processing, thus contaminated foods are lost for consumption. These losses are preventable, but at the cost of improved storage facilities and sometimes the use of pesticides and antimicrobial agents. Losses of stored foods are a major factor in the availability of food in developing countries, where not only high temperatures and humidity accentuate the potential for biological damage, but also insufficient supplies of clean water.

The storage of foods at the household level may also cause losses due to deterioration of foods. Drying, salting or cooling are effective ways of preserving foods. Drying is the removal of water and is traditionally achieved by leaving commodities in the sun. The larger the surface area, the faster the drying process. The application of vacuum and heat also accelerates the drying process. Salting in combination with smoking and drying is still an important preservation method in non-industrialized countries. In industrialized countries, some products such as smoked, salted, and dried fish or meat are still prepared, but canning, refrigeration, and freezing have become more common preservation techniques. The latter techniques have also penetrated less industrialized countries.

In principle, the only difference between industrialized and domestic food preparation is one of scale. In practice, there are differences in apparatus used in the food processing factory. Additionally, the need for control of the processes themselves has increased not only in these factories, but also in restaurants and other types of large scale catering, as has the control on hygiene as well as the raw ingredients and final products. To meet these requirements, the Hazard Analysis Critical Control Points (HACCP) was developed (Hoogland *et al.*, 1998). This is a systematic plan to identify and correct potential microbial hazards in the manufacturing and commercial use of food products. Control is required to ensure that industry and catering meet the quality standards demanded by legislation and/or by the customer.

6.3 Food distribution and consumption

Food available in a country, whether locally produced or imported, is not evenly distributed amongst people. Uneven food distribution is found between different regions of a country, rural and urban areas, between different socio-economic groups, and also between members of the household.

Supply and distribution of food in a rural household may be summarized as in Figure 6.1, which is based on Lewin's channel theory. One must find out through surveys which members of the household control the various channels: the husband, wife, or domestic staff. In some societies, the wife is responsible for buying and the husband for supplying food from the farm. Women may control the farm supply of vegetables and the men the staple foods, as in various parts of tropical Africa. In traditional Muslim societies, it is usually the men who buy food and who control the household budget.

Members of a household will not always eat together around the same "table", as occurs in most Western societies. In other parts of the world, members of the household may eat in separate groups and not all at the same time. In Indonesia, men eat first and women and children later; in parts of Africa, there are sometimes three eating groups: the men, the women and very young children, and the other children under the guidance of an older sister.

One must find out which group receives the food first, and who is responsible for apportioning the food not only as a whole but also in the various eating groups.

Apportionment of food has both a physiological and socio-cultural basis. Important socio- cultural factors influencing food apportionment are as follows:

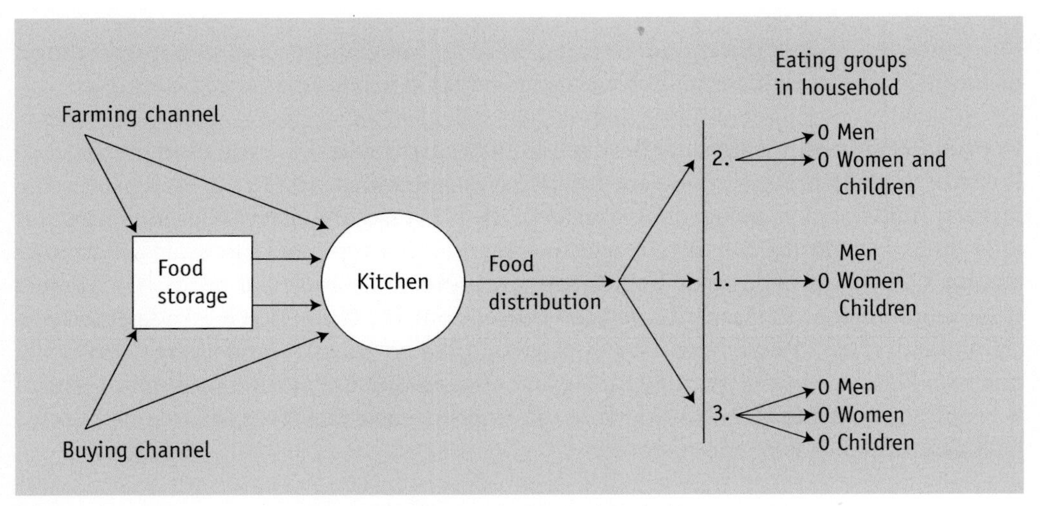

Figure 6.1. Supply and distribution of food in a rural household.

- social status in the household; *e.g.* who has the first choice of food or the sequence in which meals are served;
- prevailing concepts toward food;
- social function of food, especially food as an expression of prestige, and the obligations of hospitality.

In societies with a male dominance, it is common that meat is regarded as belonging to men to a certain extent, and it is also a means for men to reinforce prestige (Den Hartog, 1973).

Food distribution within the household is a complex system (Rogers and Schlossman, 1990). Studies carried out in Malawi and Bangladesh indicate a cultural ideology which gives food allocation priority to adult males over females. There is, however, a lack of evidence that this adds up to more than a difference to be expected on grounds of smaller body size and lower activity. In Bangladesh, it appears that young girls are less well fed than boys. This is not the case in Malawi, where girls have better access to snack foods, because boys are often absent during the day (Wheeler and Abdullah, 1988).

Food distribution in many cultures might be a consequence of a social tradition or may have a religious meaning. For instance, eating in Islam is considered to be a matter of worship of God, like praying, fasting, and other religious practices. It is said the amount of food for one person is always sufficient for two, and the amount for two is sufficient for four. It is unthinkable that a person should eat his food alone without sharing it with other people present.

6.4 Food and nutritional care

Care is increasingly recognized as an important determinant of good health and nutrition, along with food security, availability of health services, and a healthy environment. The importance of care for nutrition and child's growth was first recognized in an orphanage by Dennis (1973) in the 1960's. He observed two year old children who could barely sit up and could not speak, and who had little contact with the child care workers. As an experiment, he assigned each child to one particular caretaker who was asked to pick up, hug and talk to the child on a daily basis. The children changed radically in terms of motor development and grew well, just because they received more care (Engle and Lhotska, 1999).

Care is the provision the household and the community of time, attention, and support to meet the physical, mental, and social needs of growing children and other household members (Engle *et al.*, 1997). Care is probably the most amorphous factor among the three underlying causes of malnutrition and is difficult to measure in a concrete way (Engle *et al.*, 1997). Care refers to the behaviours and practices of caregivers (mothers, siblings, fathers, and child-care providers) to provide the food, health care, stimulation, and emotional support necessary for the health, growth, and development of vulnerable household members. These behaviours actually translate food security and health resources into, for example, a child's well being.

Care behaviours can be grouped into the major categories of:
1. care for women, such as providing appropriate rest time or increased food intake during pregnancy;
2. breast-feeding and feeding of young children, such as offering support to children to eat or encouraging a sick child to eat;
3. psychosocial stimulation of children and support to development;
4. food preparation and storage behaviours;
5. hygiene behaviours, such as hand-washing and waste disposal; and
6. care for children during illness, including diagnosis of illness and health seeking behaviours.

Improvement of the health and nutritional status often relies on behaviour: a child has to be taken to a health clinic for growth monitoring to be effective, complementary foods need to be prepared and fed to a child in order for them to be effective, *etc.* The time costs to these behaviours are often not recognized, but often add to women's already full days.

The performance of these behaviours requires enough resources for care giving, or an absence of constraints to care giving, so that the caregiver can put knowledge or expertise into practice. Resources for care include the caregiver's education and beliefs.

Several studies indicate that women with more education tend to commit more time and effort to child care than less educated women. Other resources are the care-giver's health and nutrition status, mental health, and self-confidence. Depression and stress could affect the ability of mothers to interact with their children, which can lead to failure of the child to thrive. Additionally, the caregiver's autonomy and decision-making in the home, her workload or time constraints, and finally, the social support available to her for alternate care-giving may affect the nutrition of children. Knowledge, beliefs, and education can be seen as the core capacity of the caregiver to provide appropriate care, while the health and nutrition status, and mental health (including lack of stress and depression), represent individual-level factors that facilitate the translation of capacity to behaviour. Finally, autonomy, time availability, and social support are family- and community-level variables that facilitate this translation (Engle *et al.*, 1997).

However, the care giver's behaviour is not the only important factor, but also the behaviour of the child and the characteristics of the child's environment. Children do play a significant role in determining the care that they receive. Some studies suggest that better nourished, larger children may receive more care, and that in some cultures, a poorly nourished child may be assumed to have no will to live and be allowed to die. The healthiness, perceived vulnerability, perceived weight, and even physical attractiveness of children affect the practices of their caregivers (Engle *et al.*, 1997). The affective relationship between the caregiver and the child forms the central part of this process, and problems in this relationship may contribute to child malnutrition or ill health.

It is generally observed that there is a wide dispersion of child nutrition outcomes in households with similar levels of income and socio-economic status. Frequently, even poor households can have a successful growing and well-nourished child, while rich households show failure in child growth. A large part of this variance can be attributed to differences in care behaviours. In order to derive such culturally endogenous, yet successful, behaviours that may be of relevance for nutrition interventions, the concept of "nutritional positive deviance" was introduced by Wishak and Van der Vynckt in 1976. For the most part, positive deviance refers to children who do not show evidence of malnutrition when many others living in similar unfavourable environment are malnourished (Shekar *et al.*, 1991). These children could provide examples of successful child care behaviour and supporting systems within the household and community that may provide guidance when designing programmes in these communities.

The concept of positive deviance was put into two different uses in the Hearth nutrition programs (Wollinka *et al.*, 1997). First, it was used to discover affordable and nourishing local foods that mothers can give to their children. In Vietnam, for example, this was shrimp, and in Haiti it was a mixture of beans, vegetables, and grains. Secondly, positive deviance was used as a communication method to convince mothers of malnourished children that an affordable solution existed. The meals mothers cook and feed their children are based upon information gathered from mothers of well-nourished children in their own community, the positive deviant mothers. Positive deviance is a quick, low cost method to identify the strategies used by these people and encourages the rest of the community to adopt them (Marsh *et al.*, 2004).

6.5 HIV/AIDS and its impact on nutrition and food security

HIV (Human Immunodeficiency Virus) was first identified in 1981. By the end of 2005, the disease was estimated to affect about 40 million people worldwide and be responsible for more then 25 million deaths since it was first recognized. Africa remains the continent by far the most affected by HIV/AIDS. Of 40 million people living with HIV/AIDS worldwide, 25.8 million are in sub-Saharan Africa. Two thirds of all people living with HIV are in sub-Saharan Africa, as are 77% of all women with HIV (WHO and UNAIDS, 2005).

HIV is only contracted through the exchange of bodily fluids and is transmitted through three primary routes (Anonymous, 2004). Having unprotected sex with a person already carrying the HIV virus (sexual transmission) accounts for 70-80% of transmissions globally. Despite the fact that it is the dominant mode of transmission, sexual transmission carries a risk of only 1%. The second way of transmission is through parental transmission. This can occur through transfusions of contaminated blood (with an estimated risk of over 90%), through sharing of non-sterilized needles among intravenous or injecting drug users (risk of transmission is 0.1%), and needle pricks among health workers. The third mode of transmission is through vertical transmission from mother to child. This transmission can take place in utero during pregnancy, during labour and delivery, and through breast-feeding. A pregnant woman who is HIV-infected has about a 14-50% risk of infecting her baby with HIV in the absence of antiretrovirals. Among children who become infected, 25-40% will contract HIV during pregnancy or during labour, and 15% through breast-feeding.

The link between nutrition, food security, livelihoods, and HIV/AIDS, and the devastating confluence of AIDS and food scarcity, is widely recognized (Gillespie and Kadiyala, 2005). The average life expectancy in sub-Saharan Africa is now 47 years, compared with an estimated 62 years without AIDS. The under-five mortality has also increased by 20-40% in the region. AIDS kills the most productive and reproductively active members of a household, increasing the number of dependent household members and also the number of orphans and child-headed households. AIDS mostly attacks people in the 15-50 age group, the age when people have families. Therefore, large numbers of orphans are left behind when AIDS victims die. A study in Zambia found that 68% of rural orphans were not enrolled in school, compared to 48% of non-orphans. Many children lose their parents before learning basic agricultural skills and nutrition or health knowledge.

Reduction in agricultural work or even abandonment of farms is likely. With fewer people, households farm smaller plots of land or resort to less labour-intensive subsistence crops, which often have lower nutritional or market value. A household's ability to buy food is reduced because of impoverishment due to the loss of productive family members and household assets. A study in Tanzania found that per capita food consumption decreased by 15% after the death of an adult in the poorest households.

HIV/AIDS strains the already stretched health systems and budgets. During early HIV infection, the demand is mostly for primary care and out-patient services. Demand for hospitalization increases as the infection progresses to AIDS. The increased patient load and the toll the epidemic takes on health workers lead to staff shortages and staff burnout.

A decline in school enrollment is one of the most visible effects of HIV/AIDS. Children are removed from school to care for sick parents and family members. Households become unable to afford school fees and other expenses and, in addition, children also become infected and do not survive through the years of schooling.

Women in sub-Saharan Africa are disproportionately affected by HIV/AIDS. The main reasons why they are more vulnerable than men are due to gender inequality and women's social, cultural, economic, and biological vulnerability (Loevinsohn and Gillespie, 2003). Men, due to migrant work and patriarchal ideals, often pursue multiple relationships, which increases the risk for women to be infected with HIV. AIDS worsens existing gender-based differences in access to land and other resources. In Africa, some traditional mechanisms to ensure a widow's access to land contribute to the spread of AIDS, such as the custom that obliges a widow to marry her late husband's brother. HIV/AIDS adds to women's workload as women are the traditional care-givers to the sick and orphaned children, and as they play a key-role for achieving nutrition security and fighting poverty, this further weakens existing social support networks.

The HIV/AIDS epidemic is occurring in populations where malnutrition is already pandemic (WHO, 2003b). Malnutrition is both a consequence and a contributing factor to HIV and the disease progression (Piwoz and Preble, 2000). HIV affects nutrition by decreasing food consumption, as many patients have difficulties swallowing due to painful sores in the mouth and throat. Patients also lose their appetites and suffer from medication side effects, including nausea, diarrhea, vomiting, and abdominal cramps. HIV/AIDS impairs nutrient absorption and causes changes in metabolism, leading to anorexia and fever, but also to increased energy requirements. HIV is associated with wasting as a consequence of the inability of HIV-infected persons to preserve or regain lean tissue.

On the other hand, malnutrition also contributes to the HIV disease progression and mortality (Tomkins, 2005). It is recognized that adequate nutrition cannot cure HIV infection, but it is still essential to maintain a person's immune system, to sustain healthy levels of physical activity, and for an optimal quality of life. Adequate nutrition is also necessary to ensure optimal benefits from the use of antiretroviral treatment, which is essential in prolonging the lives of HIV-infected people and preventing transmission of HIV from mother to child. However, the WHO clearly warns about the proliferation of unproven diets and dietary therapies in the marketplace which exploit fears, raise false hopes, and further the impoverishment of those infected and affected by HIV/AIDS[1].

The debate continues as to whether HIV-infected mothers should be advised to exclusively breast feed, to use breast milk substitutes, or to use alternative techniques (Tomkins, 2005). In environments where the water supply, personal hygiene, and sanitation are adequate and regular, sufficient quantities of breast milk substitutes are available, then the reduction of infant deaths from post-natal HIV transmission by using breast milk substitutes may be greater than the number of deaths from infectious diseases due to taking breast milk substitutes. However, for the majority of the populations in resource-limited countries, the number of deaths from failing to protect infants against diarrhea, pneumonia, septicaemia, and malnutrition by using breast milk are likely to exceed the number of deaths from HIV infection, and therefore exclusive breast-feeding is advised. Women are advised to follow UNICEF/WHO/UNAIDS guidelines, "where replacement feeding is acceptable, feasible, affordable, sustainable and safe, avoidance of all breast-feeding by HIV infected mothers is recommended, otherwise exclusive breast-feeding is recommended during the first 6 months of life" (WHO, 2003a).

[1] For detailed guidelines on nutrient requirements for people living with HIV/AIDS, please refer to WHO, 2003b.

7. Nutrition policy and programmes

Nutrition policies and programmes are designed to reduce malnutrition in populations. Malnutrition encompasses both undernutrition (due to some combination of food, care, and health deprivation) and overnutrition (due to a combination of excess consumption of some diet components and too little physical exercise) (Haddad and Geisller, 2005). The most common approach taken by governements to address issues confronting their countries is termed "planned development". Nutrition planning also comprises the implementation of activities through a structured framework of objectives and programmes involving the allocation of resources in a pre-determined fashion within a given time frame. A theoretical framework of the planned development process is given in Figure 7.1 and illustrates the stages considered to be most relevant (Quinn, 1994).

In order for nutrition problems to be addressed by the planned development process, the authorities in charge must first consider them to be important enough to warrant such attention. They must be convinced that malnutrition matters. There is adequate justification for why malnutrition needs to be eradicated. Eliminating the problem of malnutrition and the associated human suffering has a high ethical justification. This justification is often used by organizations to raise funds from the general population. By showing pictures of seriously malnourished children on television or in newspapers in industrialized countries, for example, it is hoped that people are appalled and donate money out of ethical considerations of the unacceptable situation. There has been a growing number of people and organizations in recent years that see nutrition as a basic human right. Additionally, malnutrition is associated with a growing feeling of insecurity in the world, where malnutrition could contribute to the hopeless situation of the poor and cause civil unrest, which may lead to an armed uprising.

Child malnutrition also has many functional consequences as it increases the risk of morbidity and mortality. Malnutrition has serious economic consequences not only for the individual affected, but also for the family, community, and nation. Poor nutrition impairs cognitive development and reduces school performance, and work capacity and labour productivity in adults are affected as well. Economic costs increase not only due to lower worker productivity, but also due to the higher costs of health care and social programmes required for dealing with malnutrition. It has been agreed that the causes of malnutrition are many and deeply rooted in socio-economic, agricultural, and environmental conditions. Survival under conditions of poverty ultimately leads to the high rates of inadequate growth found in young children and measurable deficits in height and weight. Hence, the nutritional status of young children in developing countries provides a good indicator of a country's overall socio-economic development.

This is illustrated by the inclusion of an anthropometric indicator (underweight) as one of the two indicators for the first Millenium Development Goal as formulated at the turn into the 21st millenium.

Nutrition is not always seen as a development priority in many countries, however. For example, a study by Aldana (2004) indicated that only two out of ten Poverty Reduction Strategy Papers (PRSPs) studied state nutrition as a development priority. The decision of whether nutrition is a priority is closely tied to the general perception of the problem of nutrition, which is based on a combination of scientific knowledge, ordinary knowledge, and political factors (Quinn, 1994). Scientific knowledge based on empirical data forms a central part of the general perception. Ordinary knowledge is also very important and is based on a combination of common sense, scientific facts, speculation, and beliefs. Ordinary knowledge might be considered fallible, but its influence on the general perception of certain issues may at times be immense, since it can exert considerable influence on personal and public opinion. In this respect, the role of a free press must be mentioned as reports on disasters induce governments to act. Besley and Burgess (2000), for example, found a direct link between press coverage of local crises and state government responsiveness in India. The political environment is also important and may be a decisive factor in determining whether nutrition is considered an area for action by the authorities. The decision that nutrition should be addressed by the planning machinery will be reflected in an official statement that nutrition is one of the development priorities. Where nutrition is not identified as a development priority, general opinion could be influenced through advocacy strategies before or together with improving the way nutrition is incorporated into the planned development process.

Once the decision has been made that nutrition is a national problem that needs to be ranked among other development priorities, the process of planned development can best be described as an ongoing policy cycle. In its simplest form, it is illustrated by a Triple A Approach, representing a continuous process of assessment, analysis, and action. The underlying philosophy is that an adequate assessment and analysis of the nutrition situation is necessary for the design of appropriate actions. The *assessment* of malnutrition involves the compilation of all available data and information needed to address the extent and distribution of the malnutrition problem, as well as understanding the underlying causes and how these can be addressed. If data appear to be incomplete or out of date, new data might need to be collected. *Analysis* leads to a conceptual understanding of the problem and will generate a number of potential strategies and actions, of which the "best" alternatives have to be choosen, taking into account net social benefits, administrative feasibility, and any constraints or side effects that may be relevant to implementation. In this step the commitment is shown through the allocation of resources, such as the manpower and money required to implement the proposed actions. *Action* involves the mobilization of resources and the carrying out of plans according to a time schedule. The handing over of the project and ensuring its continuation has to be included. Finally, *Re-assessment* or *Monitoring and Evaluation* is an essential part of the process, in which a comparison is made between the actual activities and outcomes with the intended ones. Lessons learned from such evaluations and monitoring exercises will determine the subsequent setting of priorities and planning as well as the revision of objectives and the refinement of programme strategy in order to maximise future impact.

Although the approach described above and shown in Figure 6.1 appears to be logical and straightforward, experience has shown that the transformation of paper plans to real life has proven to be fraught with problems and weak links in the theoretical chain of events. Problems stem from a wide range of causes, from conceptual issues, management, manpower, money, and politics. A major part of the difficulties found arises from the lack of recognition and attention given to fulfilling certain requirements. Quinn (1994) identifies four main prerequisites that need to be fulfilled. The *first prerequisite* refers to the *formulation of objectives* which are consistent, realistic, and acceptable to those in charge of carrying out the planned development process, as well as those who they will eventually affect. If there is no objective or set of objectives agreed upon at the onset, it is difficult to focus efforts towards a common goal. The objectives should also be realistic so as not to falsely raise expectations. For example, it is not realistic to expect the level of malnutrition to reduce by 10% per year. Data have shown that, due to income growth and immunization, nutrition levels improve at 0.5% per year. Succesful nutrition programmes such as in Tanzania or Thailand show a maximum reduction in malnutrition of 2-4% per year. Vitamin A supplementation may reduce mortality among under-six year olds by 23%. Additionally, the stated objectives must be specific enough so that it is understood by all what needs to be achieved, how much it needs to be improved, for whom it needs to be achieved (women, children, or the whole population), when it needs to be achieved by, and where it needs to be achieved. An example of a clear objective is: to improve exclusive breast-feeding rate (*what*) from 37% to 70% (*how much*) among children under five years of age (*who*) in northern Ghana (*where*) from 2005 to 2010 (*when*).

A *second important prerequisite* refers *to the political will to achieve the objectives*. Although it may be politically acceptable to design policies and programmes aimed at reducing poverty and malnutrition, the lack of sincere political will as well as a lack of resources may prevent the actual implementation of these policies and programmes. Assessing political will is not straightforward since statements on paper may serve to create a politically correct external image, for example, in the case a government telling the donors what they want to hear while not really meaning it. A better measure of political will is the allocation of resources to certain sectors and programmes.

The *third prerequisite* is that the *planning issues must be thoroughly understood*, including its underlying causal processes. This could be the general perception of child malnutrition and its underlying determinants and how these compare with the internationally accepted model of child malnutrition in developing countries and its underlying causes as described earlier in chapter.

The *fourth prerequisite* is related to the *existence of the necessary means and capacity* to achieve the stated objectives. Having the trained manpower, equipment, money, and time is necessary to implement the policies and programmes which have been proposed. Related to this fourth prerequisite is the issue of management as a critical issue in all development programmes. Management in general, but also in nutrition, is frequently overlooked by planners but received the previous years more attention as it forms a large part of the problem behind the failures seen in the development and nutrition planning made in developing countries. Good governance is internationally recognized as the key to success as it strengthens the incentives for investment, reduces the incentives for misallocation and misappropriation of resources, typically seeks representative democracy, does not generate conflict and, in general, advances the interests of the poorest (Frankenberger *et al.*, 2002). It should be noted that "good governance" has been subject to widely different interpretations, but a number of core principles include: ensuring transparency and access to information in all public affairs; securing the participation of civil society in the planning, budgetting, and monitoring of development processes affecting people's lives; respecting the rule of law; and the possibility of holding states accountable for their responsibilities and promises (SCN, 2004).

Much is known about how to combat the different forms of malnutrition. In general, the intervention can be divided into direct and indirect interventions (Haddad and Geisller, 2005). A number of direct interventions exist that have shown to be efficacious, effective, and cost-effective in reducing undernutrition. They exist for all stages of the life-cycle and for specific nutrient deficiencies, such as vitamin A, iron, and iodine. There is little information on the efficacy of approaches to obesity and diet-related chronic diseases in a developing country setting. Indirect interventions do not have nutrition as a primary or even secondary goal, but can have indirect and supportive effects on nutritional status. This heterogeneity of interventions is related to agriculture, income generation, education, access to clean water, sanitation, health services, *etc.* Less evidence is available on impact on nutrition for these interventions. However, it is increasingly recognized that insertion of direct intervention components into indirect interventions can be very effective. For a more extensive overview of direct and indirect interventions, refer to Haddad and Geissler (2005).

Monitoring and evaluation is an essential part of the planning process. As indicated in Figure 7.1, it forms the fuel of the Triple A approach (Assessment, Analysis, and

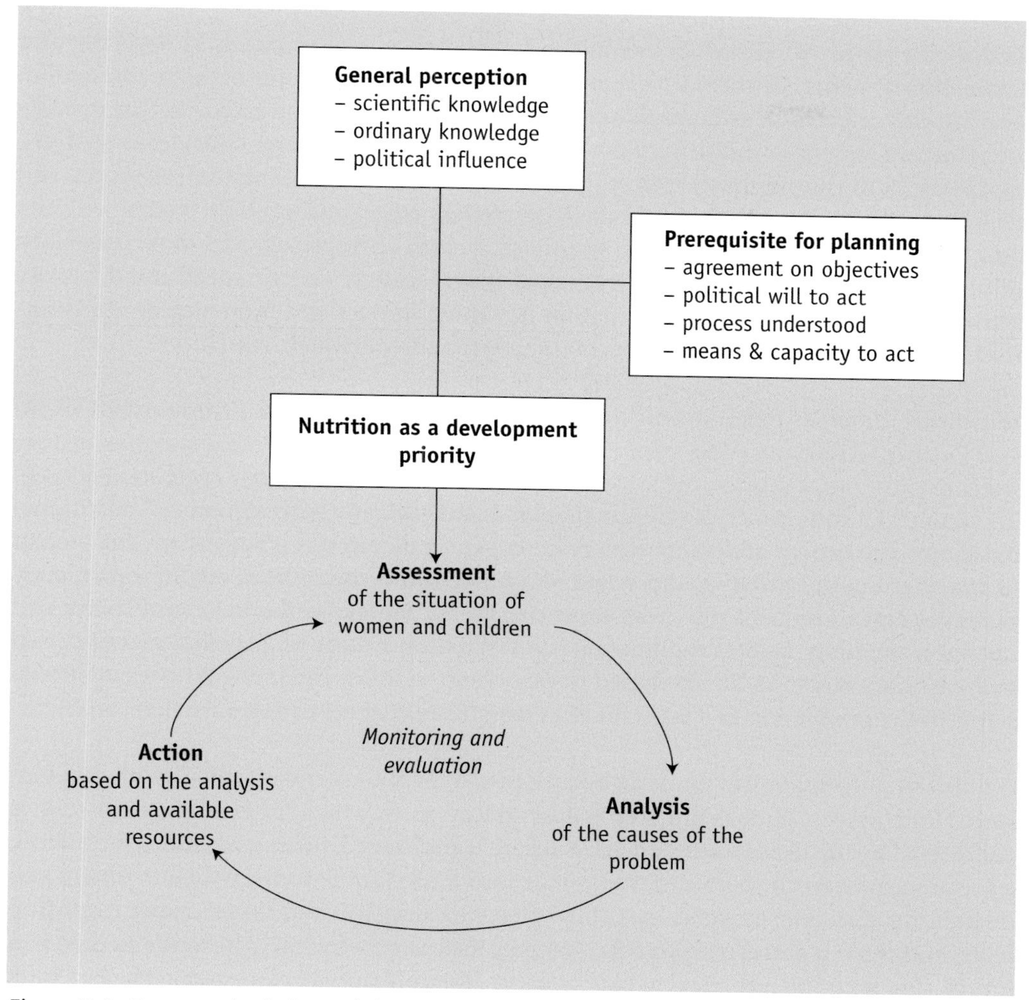

Figure 7.1. Framework of planned development cycle (adapted from Quinn, 1994).

Action). The two words originate from Latin, where "monitor" is derived from the word meaning *to warn*, and "evaluation" stems from the word *value*. Monitoring generally refers to observing or checking on research or programme activities and their context, results, and impact. Its goals are to ensure that inputs, work schedules, and outputs are proceeding according to plan, to provide a record of input use, activities, and results, and to warn of deviations from initial goals and expected outcomes. Evaluation is judging, appraising, or determining the worth, value, or quality of research or programmes. This is done in terms of relevance, effectiveness, efficiency, and impact[2]. Monitoring and evaluation systems are meant to provide information for decision making and action, and are also referred to as management information systems. The activities that provide information for the monitoring and evaluation systems (*i.e.* data collection, analysis, reporting, and distribution) should be as simple as possible. Simple procedures and techniques for data collection and analysis are less costly and provide results within a short period of time. In spite of the simplicity of such an approach and the relative lack of accurate data, a compromise between quality and accuracy versus speed and relevance must often be made. Simple information has more impact and influence on decision-makers as they are not always "experts" in nutrition (Rodriquez, 2001).

In general, different types of nutrition information systems can be distinguished (Babu and Pinstrup Andersen, 1994). For short-term planning in rationalizing and maximizing the impact of programmes, the system of *programme monitoring and evaluation* is used. To undertake long-term development plans and policies with enhanced nutritional outcomes, the system of *development planning and policy design* is suggested. The system of *timely (or early) warning and intervention* is used to meet the needs for emergency planning and to prevent the short-term critical reductions in the food availability and nutrition situation. Later, a monitoring and evaluation system for problem identification and advocacy is also to be developed to assess and monitor the indicators of nutritional status of the population as a basis for allocating resources towards a particular problem.

A nutrition information system can be seen as a "dynamic ascendant spiral" with different steps (Rodriquez, 2001). A first step is data collection, in which the raw data to be used is collected. During the second step, the data are transformed/processed into information, a meaningful transformation putting facts into a relevant context, which is timely and reliable for decision makers. In a third step, the decision should be made regarding interventions (*i.e.* activities, projects, programmes, *etc.*), which will be implemented in a fourth step with the intention to produce the desirable impact expected in the last step.

Rodriquez (2001) states that the design of a nutrition information system should comprise the following steps:

- Identify the problem and desired impact, which is essential to design a realistic system because it implies that the system will be targeted to deal with the problems assessed;
- Identify the potential policies, interventions, and programmes that are relevant to the problems;

[2] Relevance refers to the appropriateness and importance of the goals and objectives in relation to assessed needs. Effectiveness refers to the degree to which goals have been achieved. Efficiency refers to the cost-effectiveness of activities, and impact refers to the broad, long-term effects of research/programme Horton *et al.*, 1993.

- Consider the decisions that need to be made regarding the policies, interventions, and programmes. This step includes the identification of the relevant decision makers. The most frequent decisions that have to be made are related to the allocation of resources, the design, redesign, or even discontinuation of programmes and the set up of pilot programmes;
- Specify information needs for such decisions;
- Define data to be collected to provide this information. A number of practical issues should be considered: What is the quality of data needed and is this realistically obtainable? Are there existing sources of data or are new sources necessary? What is the frequency of data collection to provide timely information? What are the present institutional arrangements for data collection, analysis and reporting?

One of the major pitfalls of a management information system is that the system is unable to go beyond data collection, where the results of the system are not actually used in decision-making. Major criticisms of currently operating systems are, amongst others: too much data are collected while little is analysed and much less is reported; the time gap between collection and analysis, and between analysis and reporting is too wide so that the information comes to the decision makers too late; the information given does not match the needs of the decision makers; a lack of capacity to convert information into decisions and decisions into action. To avoid the efforts made in data collection and analysis becoming futile, it is important to design the system in continuous collaboration with the decision-makers at all stages. To increase the likelihood of the use of information in decisions and the timely implementation of interventions, it is important to organize workshops and other advocacy activities to sensitize decision makers towards understanding food security and nutrition issues and the importance of information-based decisions that reduce food security and malnutrition (Babu and Pinstrup Andersen, 1994).

PART TWO:

FIELD STUDIES

Source: FAO, 1974.

8. Before starting: Ethical considerations of field studies

Before planning and implementing field studies, ethical considerations should be taken into account. The ethical considerations deal with the integrity of investigators and the protection of the target population and community or setting where the fieldwork will be conducted. Participants and officials or institutes responsible for nutrition in the community should be aware of the project and well informed about the purpose, expectations, and practical consequences for themselves.

Approval by an (local) ethical committee is currently desired and even required to get the project funded or the results published in a scientific journal. Therefore, the principle investigator must submit a research protocol to the committee. If such a committee is not available for the research area, an ad hoc committee must be set up. In addition to the protocol, the committee will ask for the information presented to the study participants before the latter give their informed consent.

Briefly, the committee only will approve the study if the results will lead to new, scientifically sound information. In addition, the type of research must be the most efficient way to obtain the information. The information provided to the participants should also be complete and fully understood by them. Below, we will explain in more detail some of the points in the protocol which are specifically important to the ethical committee.

Aim of the study

The investigators must convince the committee that the study leads to new information. However, "new" information also means the confirmation of existing knowledge under other circumstances, or negative outcomes. Additionally, it should be made clear from the beginning whether the participants will directly or indirectly profit from the study and how this will be communicated to them.

Design of the study and methodology

The methods and design of the study should be scientifically sound, which is fully discussed in this manual. However, the study should also be practically feasible. This means that in cases of a observational field studies, skilled staff, equipment, sufficient time for observations, and especially, sufficient finances to conduct and report the study should be available.

Setting/community

Key persons in a region, setting, or community may have specific interests in the results and may be needed for the further implementation of recommendations based on the results. Therefore, a plan for the involvement of key persons in the research area is often requested. Key persons are, for example, health officers, community representatives, or teachers.

Target group/subjects

Ethical problems may arise when deciding on the information to be given to the subjects. It is possible that behaviour during observation or answers on questionnaires may be affected if the exact purpose of the study is explained to the subjects. The researcher will usually limit the information given to a more global purpose. An ethical committee will agree to this approach if the problem is adequately handled when presenting the results.

Observational data should always be treated anonymously and the researcher should always obtain the informed consent of the participant. This might be a problem when working with illiterate or mentally handicapped people, and a contact person may sign for the latter. When children are the target group, the parents should also consent to the study.

Incentives are sometimes given for participation in some studies. The incentives should fit the culture of the population and not affect the results of the study. Subjects also should know if they will profit directly or indirectly from the study. Thus, if the purpose of the study is to describe the nutritional status of the population and the results show a high prevalence of malnutrition, what will the policy of the researchers be in response?

Adverse affects / medical problems

Adverse effects are mostly a problem of trials and generally do not occur in field studies. However, if an accident does happen it should be reported to the ethical committee. A researcher should define from the outset what is to be done if the study observes health threatening symptoms in the subjects and what can be treated, such as a case of kwashiorkor.

Research staff

Research staff should be skilled and trained in the necessary parts of the study. Local staff can be very helpful because they know the culture and the food habits of the community.

However, it is possible that this may interfere with the anonymity of the data collection and may lead to behaviour or answers on questionnaires that are not authentic.

Presentation of results

Results will normally be available in two forms:
1. a project report for the benefit of the staff, the sponsors, or management of the research institute and for the community
2. in cases of contract research, a confidential research report may be written for the contractor
3. a publication in a professional or scientific journal

It is important that sponsors have no influence on the content of publications. See the following for details:

> Uniform Requirements for Manuscripts Submitted to Biomedical Journals: Writing and Editing for Biomedical Publication. International Committee of Medical Journal Editors. October 2004. Available at http//www.icmje.org/index.html (Anonymous, 2006).

Last but not least, it should be clear from the beginning which results and in what form they will be presented to the participants of the study.

9. Food ethnography

9.1 The use of food ethnography in nutrition studies

General

It is obvious that policy makers, programme staff, and fieldworkers should have data on the nutritional nature of the problem they are facing at their disposal when working with food and nutrition issues. Of equal importance is gaining knowledge and insight into how people deal with their food. Since the 1930's, several anthropologists and, to a lesser extent, geographers have given attention to food culture, food habits, and food ethnography[3].

The International Commission on the Anthropology of Food and Nutrition (ICAF) was launched in 1978 under the leadership of Igor de Garine[4]. ICAF has organized a number of workshops by bringing together specialists from the field of anthropology, ethnography, and the nutritional sciences (see *e.g.* De Garine and De Garine, 2001). Other resources which should be mentioned are a useful manual on nutritional anthropology (Pelto *et al.*, 1989); a guideline for an ethnographic protocol related to vitamin A food sources (Blum *et al.*, 1997); and a concise manual for writing a proposal on household nutrition research (Niehof, 1999).

One of the practical difficulties when working on ethnographic food data is that these data at first sight sometimes look less convincing when compared to the so-called hard data of the nutritional sciences. Or, as Audrey Richards once said about constraints within interdisciplinary food and nutrition research: "What pleases the anthropologist, displeases the nutritionist and the other way round, what pleases the nutritionist, displeases the anthropologist". One should realise that nutritional data will inform us about the nature of a nutritional problem, such as energy and nutrient deficiencies or obesity, but only partially about the causes of a nutrition problem. Food ethnography is

[3] The early pioneers in this field were the anthropologists Audrey Richards and Margaret Mead. The first collective research between an anthropologist and nutritionist was published in 1936 (Richards and Widdowson. Further should be mentioned: Guthe, C. E. and M. Mead (1945), *Manual for the study of food habits*. Of a more practical "how to do it" nature is the ethnographic manual of the French anthropologist Marcel Mauss, which contains several paragraphs on food supply and production, preparation and consumption (1947/1967). Some notes on fieldwork on food are mentioned in a manual of the Royal Anthropological Institute of Great Britain and Ireland (1954). The first report on a conference between social sciences and the nutritional sciences is of: Burgess, A. and R. F. A. Dean, Eds. (1962).

[4] ICAF functions under the auspices of the International Union of Anthropological and Ethnographic Sciences.

in a position to make a major contribution to a better understanding of the cultural and socio-economic context of a nutrition issue, its causes, and the formulation of problem-solving approaches.

The aim of the food ethnography

In essence, food ethnography treats the question of how people deal with their food and drinks (Van Liere *et al.*, 1996; De Garine and De Garine, 2001).

Food ethnography provides a descriptive analysis of the food system and food habits of a population, community, household, or a group of persons. It includes the ways in which individuals or groups choose, prepare, consume and make use of available foods in response to social, cultural and economical pressures. The obtained data are complementary to the nutritional data, making it complete. Together, these data give a more meaningful description and analysis of a nutrition problem. It is essential that a food ethnography survey is conducted when planning nutrition research or nutrition intervention programmes.

It is also important to keep in mind that being a national, or originating from the area or culture of research, it should not be taken for granted that one already has the required insight into how people deal with their food. This implies that food ethnography is not just a description, but a descriptive analysis of how people deal with their food in general or focused on an identified nutrition problem. For example, the focus might be on infant feeding and local complementary infant foods, intake of micronutrients, and the use of vegetables and fruits, household food security, *etc.* The components of an ethnographic food study centred on the food system at household and community level are presented in Figure 9.1

9.2 Methodology

An ethnographic food survey can be implemented after the research question has been defined and the type of data that will be collected has been decided upon. In other words, whether it will be a general food ethnography or a food ethnography centred on a specific nutrition question. The data will be collected by means of interviews and observation. The steps to be taken in carrying out field studies are discussed in Chapter 13. Suggestions as to the types of data to be collected are presented in Appendix A. Examples of how to construct questionnaires to be used in ethnographic food surveys are given in Appendix B. Suggestions for how the data collected on food ethnography should be presented are given in Appendix C.

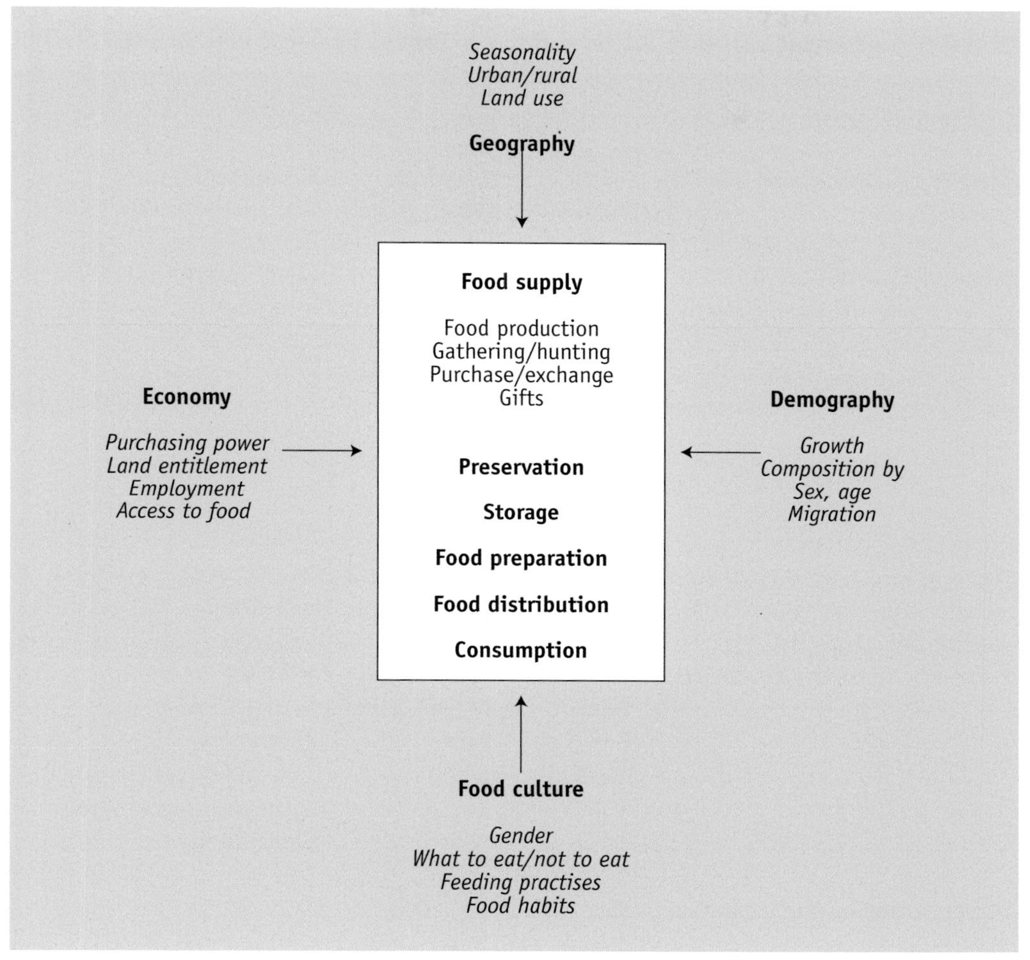

Figure 9.1. Main components of food ethnography centred on the food system at the household and community levels.

The systematic approach of an ethnographic food survey consists of three phases: Phase I, *Food Availability*; Phase II, *Utilization*; and Phase III, *Beliefs, Habits and Attributes* (see Table 9.1).

Phase I: food availability

The first phase of the food ethnography is to make a census of all foods that are consumed and available in the survey area (locally produced food and food coming from elsewhere). Regular visits to markets, shops, and other outlets, and interviews with key informants should be done. These places should be carefully visited and observed with the help of a

Table 9.1. A systematic approach: The three phases in conducting a food ethnography.

		Data	Method
Phase I	Food availability	Food and beverages in markets and shops Food crops in fields Wild food crops Domestic and wild animals Gathering and hunting Fuel wood and other energy sources Fetching water and fuel wood gathering Seasonality	Market survey Key informant interviews Observation
Phase II	Utilization	Food preparation Preparation of beverages Distribution and consumption	Observation 24-hour recall and food frequency
Phase III	Beliefs and attributes	Food habits What to eat and not to eat Norms and values concerning food	Pilot sort Food attributes and differences In depth interviews Focus group interviews Observations

Source: based on Van Liere *et al.*, 1994; Nana *et al.*, 2005.

local resource person(s). The resource person should be clearly and carefully requested to point out and to report every item that is considered to be edible by the community. The rapport between the resource person and the researcher has to be such that foods perceived as "unusual" by outsiders will be reported as well, such as insects, certain rodents, or dogs. The same applies when going into the fields looking for cultivated crops, fruit trees, and foods collected in the wild, or with domestic and wild animals.

A final list of foods will be prepared with the help of the local resource persons so that both English (French or Spanish) names together with the local names can be added on the list. The scientific names can be found by using publications on botany and zoology, and of course, with the necessary advice received from local experts in the field. An example of a list of consumed food products is given in Appendix C9.

Food availability depends greatly on the seasons. At the household level, it also depends on the means members have for access to food. Periodic food availability should be recorded for major foods. Food-related activities have a time dimension. It is necessary to construct an agricultural calendar indicating food availability and the related agricultural activities. When applicable, foods collected in the wild, food plants and animals have to be included as well. The necessary data can be collected by means of interviews from knowledgeable persons, both women and men, such as the community elder, agricultural extension agents, social and health workers, midwives, local women groups, and local leaders. In vulnerable geographical regions, much attention has to be given to the phenomenon of seasonal food shortages and famine.

Phase II: utilization

The second phase of a food ethnography deals with how people get, use, and consume their food. Careful attention should be given to preparation of daily meals and special dishes, food allocation and the eating groups within the household, and the dietary pattern. It also includes preparation tools, methods of cooking, and food storage. When working in an area or community it is very helpful to gain insight into the general dietary pattern. An example is presented in Appendix C10.

Phase III: beliefs and attributes

This phase includes identifying ideas and beliefs surrounding food, such as what can be eaten and what should not be eaten, such as food avoidances, foods for specific groups, preferred foods, and foods for special occasions. Additionally, the role of the life cycle has to be taken into account. A detailed analytical description should be made of the food supply activities, including the gender dimensions. Data can be obtained by means of interviews, group discussions with women and men, and by observation. Semi-structured questionnaires can be used to collect data on *food attributes*. Respondents will be asked to describe the perceived attributes or characteristics associated with individual foods. The *pile-sorting* method makes use of pictures (drawings or photographs) of the main foods consumed with the aim to study the relationship among a number of key foods in the community. The respondents will be asked to sort the foods that belong together to any criteria that make sense to them. Respondents will be asked for the reasons why they have sorted foods the way they have. More detailed information on the two methods is to be found in the guideline provided by Blum *et al.*, (1997).

The nature of the food ethnography is a descriptive survey of qualitative data on how people deal with their food. This does not exclude the presentation of quantitative data; to the contrary, the dietary pattern can be quantitative presented by means of the relative frequency of food consumption, where foods are obtained from, duration of breast feeding, age when complementary foods are given to young children, portion sizes, *etc.* (see Appendix C11-20).

9.3 Measuring food trends

General

The time dimension is a part of many nutrition studies, investigating issues such as trends in food consumption or trends in nutritional development, for the better or worse. Dietary surveys make use of the so-called dietary history method, with the objective being to find out the daily food consumption pattern. Respondents are asked what they ate during the preceding three days or the past week. The dietary history record method can also be used to determine the frequency of nutrition-related diseases over the past year.

The nutritional situation is never static. It is necessary to find out whether a specific nutritional issue is a recent phenomenon only, a chronic situation, or the outcome of a longstanding downward trend. Examples could be a possible decline of breast-feeding, increasing food insecurity, frequently recurring seasonal food shortages, or the insufficient availability of fuel wood for cooking. A food trend in a community can be assessed by two approaches.

- The study of available statistical material and related records, agricultural reports, health reports, and social reports of both governmental and non-governmental services and institutions. In many cases, or sometimes in most cases, this information is not available, however, so an alternative method must be found.
- For communities where no or insufficient records are available, the method of *oral history* is a very practical alternative.

Measuring trends: three-generation study

Anthropologists have developed a method called oral history for communities where no records are available. Oral history refers to the collection of any individual spoken memories concerning the respondent's life, of people he has known and events he has witnessed or participated in (Hoopes, 1979). Oral traditions, including food and preparation methods, have passed from mouth to mouth for time which reaches beyond the lifespan of the respondent. Data on past events can be collected by means of interviews.

The oral history method can be used as a research tool in ethnographic food surveys when researching food trends. This can be done by means of the *three-generation study* method using women, but not exclusively, as informants (Brouwer, 1994). A time span of 50 years can be covered by interviewing three generations:

Daughter	20 –30 years of age
Mother	40 –50 years of age
Grandmother	60 –70 years of age

Women in each category can be asked about food and related issues during a certain period of life, such as during the first years of their marriage or the first years after the birth of her first child, *etc.* In this way, the trend in infant feeding practises can be assessed.

A number of points have to be taken into account when selecting informants. One may start with the selection of women within the age range of 60-70 years old. It is possible that not all three generations are living in the same community anymore due to migration or mortality. In certain cultures, the married woman moves to the village of the father of her husband (*patrilocal residence*). In such situations one can use a more cohort-oriented approach by sampling three age cohorts from the community. These cohorts do not necessarily share family bonds. In some societies, women in their early twenties still have no major food preparation or fuel collecting responsibilities as compared to other women, such as their mother in law and other elderly women.

A methodological difficulty with the oral history method is the memory of informants and the dating of events. The use of a person's memory raises questions about validity and accuracy of the answers. Events are stored selectively and memories may be influenced by later events. Because of present experiences, events in the past can be idealized, minimized, or just exaggerated. These kinds of constraints in a three-generation study are also encountered in the dietary history methods of food consumption surveys. To help respondents recall food issues in a certain period, it is often necessary to refer to a special well-known event in that period. This can be personal life-events such as the birth of children, a wedding, or political or natural events having a direct impact on the community. By first posing some general questions, the memory of the respondents can be refreshed. The use of visual examples can be helpful for more detailed questions. If one is interested in trends in the use and consumption of wild foods collected in the fields and forest, the respondents can be shown samples of such foods. This will increase the validity of the answers given.

10. Measurement of food consumption

The concept of food consumption to be measured varies with the objectives of the survey. It is essential that a distinction is made between food consumption surveys mainly used for national food planning and administration, epidemiological surveys whose emphasis is primarily on the relation between nutrition and health, toxicological studies interested in food exposure to biochemicals, and metabolic studies focused on the fate of nutrients in the body.

Such a distinction is necessary because each objective demands different types of information. For food supply planning, one needs to know the demand and price of food and farm incomes, so data are required on the amounts of foods taken from different distribution channels by various categories of users at different times (Figure 6.1). Thus, the results of this type of survey should be expressed in terms of the amounts of foods as purchased. For economic analysis, there is less interest in what users do with food after it has been bought. If the study serves to assess the nutritional adequacy of the diet in relation to requirements or to assess food exposure to environmental chemicals, the focus will be on the amounts of food as eaten. To assess the nutritional adequacy of the diet in relation to nutritional status, results are commonly converted into amounts of energy and nutrients.

For each objective, various types of surveys exist and the approaches are at different levels:

- national accounts of annual food availability per head of population; or
- food balance sheets;
- family budget and household consumption surveys;
- individual food intake or dietary surveys.

10.1 Food balance sheets

A food balance sheet is a national account of the annual production of food, changes in food stocks, imports and exports, and the distribution of food over various uses within the country. This account can be prepared on the basis of the calendar year, the agricultural year, or the crop year. The various uses are listed under the following headings: animal feed; seed; industrial uses; waste; and the net food availability for human consumption at the retail level. Per capita food availability is given for the total population actually partaking of the food supplies during the reference period, *i.e.*, the population present within the geographical boundaries of the country at the mid-point of the reference period. In some countries, the per capita food availability refers to only the civilian population, meaning that armed forces are excluded. Per capita food availability is expressed in grams of food and is further analysed in measurements of energy and some nutrients (Cameron and Van Staveren, 1988).

Problems in collecting and handling data from food balance sheets

The FAO has compiled and published food balance sheets for most countries in the world since the late 1940's (see www.fao.org). In addition, the countries themselves or other organizations also collect these data and their results may differ considerably. The main limitations in collecting and handling these data are (Serra-Majem *et al.*, 2003)

- Food balance sheets are derived from a variety of official statistics and their quality varies between countries and commodities.
- Trade across national borders may affect results differently per commodity, per year and per country.
- Growing or catching food for home consumption and other products that do not enter the commercial market system give rise to bias in the results.
- Waste also varies between countries, between foods, and over time. These variations have to be taken into account when interpreting food balance sheets.

Problems encountered in collecting data for food balance sheets make the data of limited use to the nutritionist. However, sometimes it is the only data available on the national food supply. Imprecise data (although not inaccurate) is better than no data at all. If collected in a well defined and standardized way, a food balance sheet can give a nutritionist the following information:

- a rough indication of the adequacy and the diversity of the national food supply;
- food availability relative to other countries;
- trends in food supply if food balance sheets have been compiled over some years.

To interpret trends, one must be aware of interfering factors such as trends in different segments of the population, such as an increase of the elderly. Food balance sheets give no information at all about how the food supplies are distributed between different segments of the population. Therefore, results do not indicate which groups are vulnerable in a population.

However, food balance sheets may distinguish differences in food supply between countries. Together with morbidity and mortality data, the information is used to generate hypotheses for epidemiological studies on diet and diseases.

10.2 Budgetary and household food consumption surveys

Some main characteristics of budgetary and household surveys are as follows:
- The unit of observation and tabulation is a household or family group.
- Amounts of food purchased or (sometimes) eaten must be measured.
- Results are usually presented as averages for groups of households having a common set of characteristics.

These types of surveys do not usually consider food intake, since it may not be feasible to measure consumption outside the home. They do not indicate the distribution of food within the family and therefore are sometimes combined with individual dietary surveys.

Budgetary surveys of families

Budgetary surveys emphasise income and expenditure of food and non-food commodities. Little or no information is obtained on foods actually used in private homes. However, some countries try to exploit this method more fully, adapting it to provide particulars of amounts of food consumed as well as expenditures on food. A further expansion might classify foods and include age, sex, and occupation of the people present at each meal. These extra data make nutritional evaluations of the average food consumption of various categories of households possible. However, this additional information might overburden a budgetary survey, which is usually already heavily loaded with questions (Kelly *et al.*, 1991).

Methods used to ascertain food consumption of the household

There are at least four ways of collecting data on household food consumption (Kelly *et al.*, 1991; Wolfe and Frongillo, 2001; Trichopoulou and Naska, 2003):
1. *The food-account method* aims to record all food purchases and food brought into the household from other sources during a seven-day period or longer. This measure assumes that there have been no significant changes in household food inventories.
2. *The inventory method* aims to record acquisitions and changes in food inventory of the household, generally over a period of one week. An inventory is made of food in the house at the beginning and end of this period, and all foods brought into the house during this period are recorded as well. Thus, the only difference between this and the food account method is that changes in the food reserves or stores of individual households are taken into account. The advantage to this method is that it gives a more accurate estimate of nutrient availability than the account method. However, it distorts normal food purchasing patterns by drawing the respondent's attention to existing larder stocks. Therefore it has less value for economic purposes.

3. *The list-recall method* requires that the interviewer uses a list of major food items in a structured questionnaire to help the respondent recall the amount and price or purchase value of all foods used in the household within a specified period, usually over a period of seven days.
4. *The household-record method* aims to weigh or measure in household units the foods available for consumption in the household. This is done by the caretaker or during daily visits by field investigators. Foods not eaten by the household members should be subtracted from the total amount available for consumption. Survey subjects may individually record foods eaten outside the household. The total amounts of all the foods eaten should give a complete account of the food consumption for all subjects during the survey period.

Treatment of results

In order to make a nutritional evaluation of the food consumption or food available for the different categories of households, survey results are expressed as energy and nutrients "per household" or "per head" during a day or week. The age and/or sex structure of the family is then ignored. Comparison between households or communities will be valid only if the composition of the groups is similar. There are two other approaches possible when comparing households:

a. Instead of expressing intakes per household member, one can calculate adult equivalents (sometimes called consumption units). A drawback is that this system is based on energy requirements relative to that of an adult man. However, for different age categories, the requirements of protein, some minerals, and vitamins are not proportional to energy requirements (Table 10.1).
b. Another approach is to express energy and nutrient intake in proportions of daily intakes recommended by expert committees. By adding up the daily recommended intake of a nutrient for every household member and dividing the household supply

Table 10.1. Comparison of Consumption Unit Requirement for a reference man (body weight 62 kg) with a boy (24 kg) for energy, protein and iron. Protein requirement assumes a net protein utilization (NPU) of 60.

	Energy		Protein		Iron	
	Kcal/d	cons. unit	g/d	cons.unit	mg/d	cons.unit
Adult man	2500	1	50	1	8	1
Boy 7 year	1500	0.60	22	0.44	10	1.25

Recommendations are based on Dietary Reference Intake. Food and Nutrition Board; (2000, 2005).

by this recommended amount x 100, the percentage of the recommended supply for the household will be known:

$$\frac{\text{amount of nutrient available per day}}{\text{amount of nutrient daily recommended}} \times 100$$

In making such a comparison, food that was wasted or given to visitors or pets, or food eaten away from home should, of course, be taken into account. The main advantage of assessing dietary adequacy in this way is that it allows households of differing composition to be compared or pooled, assuming that biases in the survey apply equally for all household types. Given the inaccuracies inherent in these calculations, at least 20 families should be included in each group (Flores and Nelson, 1988).

In summary, specific points to consider in a household food consumption survey are:
1. Household food consumption surveys record food availability, not food intake.
2. The period of recording, seasonality, food purchasing patterns and sometimes salary-days should be noted.
3. Problems with non-response (chapter 13) and how to deal with it.
4. How to deal with food wasted, food given to visitors or to pets.
5. How to include food obtained outside the household.
6. Information on the distribution of foods within the household normally is not obtained.

10.3 Individual dietary survey

Several methods with various characteristics have been developed to assess food consumption (Table 10.2). The methods are often grouped into four main categories which are partly based on these characteristics:
- Recall of past intakes.
- Recording of present intake.
- Shortcut methods (*e.g.* some food frequency questionnaires).
- Combination of methods.

Recall of past intakes of individuals

Recall methods aim to elicit actual past intakes as remembered in an interview or with a questionnaire completed by respondents. The principal procedures for recalling the past intake of individuals are:
- 24-hour recall;
- dietary history.

Table 10.2. Characteristics of dietary survey methods.

Collection of data	Observation, record, interview face to face, by telephone or internet based
Time frame	Incidental diet, usual diet
Portion sizes	Weighed, estimated (models), frequency only
Conversion into nutrients	Chemical analyses, nutrient data base, food scores only

Adapted from: Van Staveren and Ocke, 2001

The 24-hour recall method aims to ascertain the food intake of an individual during the immediately preceding 24 hours or for the preceding day by means of detailed questions. Food intake is usually assessed in terms of household measures. This method estimates the food actually eaten, as recalled from memory. An interview structure with specific probes helps the respondent to remember all foods consumed throughout the day. Sometimes there is a checklist at the end of the interview with foods or snacks that might be easily forgotten. Computer assisted 24-hour recalls usually consists of multiple steps, which is especially encouraged for regions with a complex and highly varied food pattern (Slimani *et al.*, 1999; Conway *et al.*, 2003). The interview is less complicated than, for instance, the diet history method and does not take much time for either interviewer or respondent. If the procedure is restricted to one interview per respondent, information is limited to the food intake on one particular day, though day-to-day variation can be high for most people. Therefore, the 24-hour recall is often repeated or used together with another method (Section 10.3, Combination of methods). If not, the data are still useful for mean intakes of population groups.

The dietary history method (Burke, 1947) is a technique for estimating usual dietary intake. The technique is based on the premise that people have a constant daily pattern in their food habits. The method was originally developed to measure diets over a period of time for research on human growth and development. The rationale was that long-term food habits may yield clinical and laboratory signs and findings. Current intake may not reflect usual intake and so may have less value in evaluating nutritional status. The dietary history interview technique requires highly trained interviewers with a nutritional background. The data may be collected by the question, "What do you usually have for breakfast?", sometimes coupled with "what did you have for breakfast this morning?". The amounts are recorded in common household measures. The complete day is covered in this way. If an individual does not have a constant eating pattern, a dietary history cannot be compiled and thus another method should be used. The dietary history, including a checklist of foods and a cross check of all foods actually consumed in a 3-day period, may be appropriate in the assessment of nutritional status and is not a great burden for the participant. Skilled interviewers are crucial, however.

Recall methods in general have the following disadvantages. Respondents must have a good memory and (for diet history) a well-defined pattern of diet. The methods make heavy demands on the interviewer, who has to gain the confidence of the participant to make good estimates, avoid suggestion, and be capable of judging the reliability of replies. Interviews should be arranged in the house of the participant rather than in a clinic to gain the confidence of respondents, despite the extra time this may take.

Advantages of the recall method are as follows. In general, cooperation will be satisfactory because the interviews are no great burden for the participants. Dietary history and cross-checking give a picture of food intake by a group of individuals over a past period of time. Food intake can be related to other nutritional status parameters. The 24-hour recall may give a more exact picture of the actual food intake of groups. If a 24-hour recall method is repeated in the same group, it also can give information on the distribution of intakes within the group.

Recording of present food intake

Recording methods estimate the current food intake during one or more days. The amounts of food eaten can be weighed or estimated in terms of household measures.

The weighing method assesses the cooked weights of the total portions of the meal served, the portion for each individual, and leftovers. Often the ingredients and amounts used in the preparation of dishes are also measured. According to the cooperation and capacity of the participants, this method requires varying degrees of supervision. Educated people can weigh items for themselves with a spring balance provided for this purpose. With less educated people, the actual weighing should be done by the field-workers, meaning that the nutritionist has to spend several hours each day with the mother, for example, which may cause some interference in the home. The extent to which this may alter food intake is difficult to determine. A compromise must be reached between close supervision with consequent interference in the home routine and very little (perhaps inadequate) supervision so as not to upset the home pattern. An example of a weighed food record is given in Appendix D.

An estimated record is a list of all foods eaten by an individual during a specified period, given in terms of household measures or compared in size to food models. For educated people, this method is less demanding than weighing; they can record food intake themselves. There is less precision in this process but closer cooperation between fieldworker and respondent (Marr, 1971; Nelson *et al.*, 1989). Supervision by a dietician at the beginning and end of a period is necessary. A detailed interview is desirable at the end of the survey to allow for the checking of amounts. Details overlooked or omitted reduce the accuracy with which measurements can be converted to mass. For less educated people, this method is not appropriate because they cannot record and describe their portions. If a nutritionist has to do the work and spend significant time in the house with the mother, she could better weigh the foods.

Recording has the following disadvantages. It can usually be conducted for only relatively short periods (one week at most). It is a real chore for the participants; not every individual is willing or able to weigh the diet or to record the daily intake in household measures. Additionally, recording may alter the usual pattern of intake.

Advantages are that it gives a fairly exact picture of the actual food intake of a group. If weighing is continued long enough, reliable information about the food intake of an individual might be obtained.

Shortcut methods

The recall and recording methods discussed so far provide quantified measurement of daily food consumption. Sometimes less extensive and only qualitative information is required. For this purpose, a food frequency method is developed. With a short list of foods for the qualitative classification of dietary patterns, this method permits the rating or grading of items into categories so that extremes can be identified (see Appendix B). The questionnaire should consist of simple, clearly defined questions which can be used by untrained staff or can be included in questionnaires to be completed by respondents. The first food frequency questionnaires (FFQ) were developed for large epidemiological studies, such as on the relationship between diet and chronic diseases. Other methods may put too heavy a burden on subject and investigator in such studies.

Food frequency lists are also included in food ethnography (Chapter 9.2) and other approaches to measure dietary diversity (Arimond and Ruel, 2004). Dietary diversity is seen as a key component of healthy diets and, according to the literature, is associated with an increased likelihood of meeting nutrient requirements. The exact mechanism involved still needs to be elucidated. Before the FFQ can be recommended for widespread use for this purpose, additional research on the predictive value of this measurement has to be conducted.

Clearly the FFQ always has to be adapted for the purpose of the study, the target group and the food culture. The development and validation of the food frequency method is expensive. Thus, it is only cost effective when either applied in the context of a food ethnography or when diet adequacy is the purpose of the study in large-scale surveys (Cade *et al.*, 2002).

Combinations of methods

All methods have specific advantages and disadvantages. There is no best method for all purposes. Investigators should carefully consider what the best method is for their purpose. Very often, a combination of two methods might give fuller information. For instance, a combination of a weighing record at the household level and 24-hour recall for individuals can give information about food purchases for different categories of households, about recipes, and about the food intake of groups of individuals. A combination of a dietary history and current recording gives past food pattern information and a more exact picture of current food intake.

10.4 Validity and reproducibility of the methods

The quality of the measurement is determined by the validity and the reproducibility of the method. Validity is an expression of the degree to which a measurement measures what it purports to and is statistically associated with systematic error. Reproducibility is the extent to which a method produces the same results when applied repeatedly in the same situation. Reproducibility is associated with variance, comprising random response errors and true or biological variability.

To examine the validity and reproducibility of data obtained by dietary methods it is important to consider questions such as those phrased by Beaton *et al.*, (1997). What does the estimated intake really represent? What is the nature and magnitude of the error in that estimate? What is the implication of the error for interpretation of analyses?

To answer these and similar questions, methods should be validated. Two ways of validating a dietary assessment can be distinguished. The developed method may be compared with another method designed to measure the same kind of dietary data. Alternatively, the new method may be validated against some external criteria (*e.g.* a biological marker). The criteria for the validation procedure partly depend on the purpose of the study in which the method will be used.

The purpose of the study and the validity of the method

Experimental versus observational studies
It should be appreciated that there is a difference between the purposes of dietary experiments in metabolic wards and the aims of observational studies in so-called "freeliving" populations. Whereas nutritional experiments typically examine the effect

of changing nutrient intake on indices of nutritional status in a fixed period of time, population-based observations are commonly used to examine the association between the usual dietary intake of individuals or groups and other characteristics, such as disease status. There is a tendency to consider the results from metabolic studies as being more accurate than population-based observations. For the study of nutrition and health, however, both designs are useful since their purposes, limitations and advantages are different, as shown in Table 10.4.

The several-day weighed method is the preferred approach in metabolic experiments. This technique is being considered as the gold standard against which other methods are often compared. However, for observational studies in large populations or epidemiological research, this method is too cumbersome and time consuming. Thus, the 24-hour recall method has often been used instead.

Types of information required in epidemiological research

Generally, one or more of the following types of information might be sought in epidemiological research (Beaton *et al.*, 1997; Willet, 1998):

1. Assessment of the mean energy (or nutrient) intake of a group (Type 1 study).
2. Assessment of the food consumption distribution, or more specifically, the percentage of malnourished subjects in a population, without seeking to identify them individually (Type 2 study).

Table 10.4. General purposes, limitations, and advantages of an experimental and observational design for food consumption studies.

Design	General purpose	Limitation	Advantages
Experimental design	To examine the effect of nutrient intake during a fixed time period on indices of the nutritional status	Artificial conditions Limited number of subjects Limited time period	Controllable data collection intake during a fixed period of time collection
Observations in free-living subjects	To examine associations between diet and health	Data collection difficult to control Longer but still limited periods of time	Large groups and real life situations

3. Classification of individuals into extremes of the food consumption distribution, or ordering them according to quintiles or tertiles for assessment associated with some other characteristic (Type 3 study).
4. The absolute magnitude of an individual's food consumption (Type 4 study).

The last type of information refers to clinical use and will not be discussed further in this context.

Random versus systematic error

The impact of random or systematic errors on the results of the study depends on the type of information needed. Any systematic error will invalidate the results from Type 1 and Type 2 studies. However, a biased measurement might not affect the results from a Type 3 study, unless this systematic error differs among subjects and is associated with the characteristic at issue, as for instance, in the case of an energy-balance study, when only obese subjects underestimate energy intake. In most other situations, estimation of an association with the disease indicator is the ultimate goal of Type 3 studies. For these types of studies, ordering subjects correctly according to their dietary intake would be sufficient and a shift along the measurement scale may have no impact on the assessment of the effect, such as a relative risk.

On the other hand, a random error is always detrimental. There are ways to overcome this problem through special study designs, however. Therefore, information about sources of error and variability of intake is necessary. When the objective of the study is to estimate the individual mean value for a subject, the within-person variation is the relevant piece of information. The between-subject variation is also needed when a group mean is being estimated. Generally, the estimated total variance (including random error) will be available. This is the sum of within-individual and between-individual variance:

$$S^2_{\text{total}} = S^2_{\text{within}} + S^2_{\text{between}}$$

In Type 1 studies, the total measurement error will affect the precision of the estimation of the group mean to a slight extent, and this may be outweighed by increasing the sample size of the study group. The between-subject variation in a Type 2 study will always be overestimated by the total variance if single dietary assessments are made. This is due to the fact that the within-person variation is unknown. However, the within-subject variation may be estimated from duplicate observations and adjusted for in the analysis.

In Type 3 studies, attenuated associations (biased towards the null) will result from any misclassification due to random errors. The amount of misclassification, however, may be reduced by performing repeated observations on each subject, thus improving the effect estimate. In addition, as many investigators (Thompson and Byers, 1994; Beaton *et al.*, 1997; Kaaks *et al.*, 2002) have indicated, the magnitude of the unbiased measure of association can be estimated if within-subject variance and between-subject variance (or simply their ratio) are known. So, whereas the investigator should always be concerned about the impact of systematic errors, random errors should be less of a concern if the design of the study makes it possible to account for within-subject variation. Depending on the nutrient at issue, many replications may be needed due to high day-to-day variations in intake.

Note that the within-subject variation in the dietary history method does not contain the day-to-day variation, since this method examines the usual diet in one interview. Consequently, random error is smaller and, thus, reproducibility is better. However, the validity of this method has often been questioned since the individual has to remember how frequently many different items of diet are eaten.

Several studies validated the dietary history against records of food intakes. In most of these studies, data obtained by the dietary history seemed to be an overestimation when compared with data obtained from food records (Cameron and Van Staveren, 1988; Visser *et al.*, 1995).

A rough indication about the underestimation of energy intake on a group level can be obtained by calculating a resting metabolic rate based on body height and weight, sex, and age. FAO/WHO/UNU (2004) research indicated that the energy intake of populations in energy balance should be at least 1.4 x resting metabolic rate by light physical activity levels.

Subject-associated bias

As has been stated for Type 3 studies, a systematic error is a matter of concern if it varies among subjects. It has been suggested earlier that systematic underestimation may in fact be associated with factors such as obesity. Also, the so-called flat slope syndrome (phenomenon of "talking a good diet") is a manifestation of bias that varies between subjects. Such a bias of individual subjects does not necessarily have to be associated with the outcome variable at issue. For instance, in a follow-up study on the association between food intake and cancer, bias in the assessment of energy intake can be expected to be similar in future cases and non-cases. In such a situation, subject-associated bias will present itself in the data as disturbances due to the sampling procedure and will thus be interpreted as random error.

10.5 Evaluation of dietary intakes

Food intake is usually evaluated by comparison with recommended intakes of nutrients and energy. It is important to remember, however, that dietary intake figures alone can never prove whether people are adequately fed. In addition, data on the nutritional status are required. Estimates of height and weight, for instance, should be included in all dietary surveys. A simple comparison of intake levels with recommended dietary intakes or allowances (RDI/RDA) as summarized in Appendix E inevitably overstates the real problem. Recommended intakes were originally developed for planning purposes and most often they are based on minimum requirements for the group, plus two standard deviations or plus a fixed percentage. The USA National Institute of Health developed the concept of Dietary Reference Intake (DRI's) in 1998, providing reference values that are quantitative estimates of nutrient intakes to be used for planning and evaluating diets for healthy people. The DRI's include the above-mentioned RDI; Adequate Intakes (AI), which when RDI are difficult to define; Estimated Average Requirement (EAR), sufficient for half the healthy population; and the tolerable upper intake level (UL), the highest level of daily intake of a nutrient without risk of an adverse health effect. The publications of the IOM (Institute of Medicine, US) explain how to use these Dietary Reference Intake data. It is beyond this guide to explain the concepts in detail, but a short summary is given in Appendix E, Table E6.

10.6 Conclusion

In conclusion, the following remarks have to be born in mind when making a choice for the dietary method to be used.

The appropriateness of a dietary assessment method depends on the purpose of the study. This means that a distinction should be made between socio-economic studies mainly interested in foods as purchased or as eaten, and studies focused on the relation between nutrition and health, mainly concerning food and nutrient intake.

The type of information needed, the target group, and the time frame of the study must be formulated as precisely as possible. Consider whether the results of the study have to be compared with other data already available or with dietary reference intakes, as this may have consequences for the study design. Consult a statistician, beginning with the design of your study. The use of food composition tables also has to be determined in an early phase of the study. The next chapter will discuss the latter topic.

11. Conversion of amounts of foods into nutrients

When the daily amounts of foods consumed are recorded, the next step will be the conversion of these data into amounts of nutrients. Energy and nutrients in the diet may be calculated using tables of food composition or can be chemically determined from duplicate portions, aliquots, or by the equivalent composite technique. The method by which conversion will be done has to be decided upon early in the planning stage of the survey.

11.1 Using food analysis

Analysis of duplicate portions

If the consumed food has been weighed, chemical analysis of a duplicate portion is accurate. Therefore duplicates, equalling exactly the same as all the foods eaten during a 24-hour period, should be collected in a container. However, this method is not absolute, since the duplicate analysed is not necessarily identical to the portion eaten. Small items taken with or between meals might be overlooked. Furthermore, the duplicate might be prepared differently, so that the subject perhaps eats less or dishes which are different from what he or she would normally eat (Pekkarinen, 1970).

Specific problems exist for food eaten away from home. Ask people to buy a duplicate portion and reimburse them for the cost of this extra food. In addition, it is unusual to collect drinks with high water content, such as alcoholic drinks, tea, coffee, or soft drinks. These drinks highly dilute the contents of the container, making analysis less accurate. Ask subjects to weigh and record these drinks. The reader is referred to Appendix D4 for detailed instructions. The analysis of duplicate portions is time-consuming, very expensive, and can be used only in a study with highly motivated subjects. Consequently, it is difficult to randomly sample a population for such a study. The accuracy of results is also influenced by errors due to chemical techniques. Although these errors are presumably relatively small compared to the sources of error in calculations from food tables, even the most accurate method is not absolute. In general, this approach is too expensive for most field surveys. However, it might be used as a check on the database in a sub-sample of the population.

Analysis of aliquots

All foods eaten during the survey are weighed and all beverages drunk are measured, and all aliquots (*e.g.* a tenth of all foods and beverages, except for water, consumed) are collected daily. The bulked aliquots of the survey period are subsequently chemically analysed. This approach is probably less accurate than the analysis of duplicate portions because the aliquot samples are not necessarily identical to the foods eaten in mixed dishes.

Equivalent composite technique

The equivalent composite technique consists of two parts. All consumed foods and drinks are weighed or measured and recorded throughout the duration of the survey. Afterwards, a sample of raw food equal to the mean daily intake by an individual during the survey period is taken for chemical analysis. Aliquot sampling can be considered to be more accurate, however, as additional sources of variation are introduced. First, unprepared raw foods, rather than prepared foods, are most often used for chemical analysis. Second, the foods analysed may differ qualitatively and quantitatively from the foods eaten. The approach brings organizational problems, since food samples can be composed only after dietary intakes have been calculated. However, this method also has advantages. The food samples are easy to collect and the approach is cheaper than the duplicate portions or aliquot samples (Pekkarinen, 1970). It is also applicable where individual cooking and eating practices do not affect results, such as in the assessment of dietary fibre, for example.

11.2 Using nutrient databases or food composition tables

Nutrient databases list the nutrient and energy content of foods, presented as food composition tables or as computerized databases (*e.g.* www.nal.usda.gov/fnic/foodcom). Computerized databases have a substantial advantage over printed food composition tables, as they can contain a greater volume of information and the data can be used in calculations much more easily (Greenfield *et al.*, 2003). In small-scale community studies, however, the printed tables can still be very helpful. Many composition tables only give the nutrient content of raw foods, but others include prepared dishes. Those who prepare food tables have difficulty in ensuring that the foods listed are representative of the samples. Difficulties arise even with unprepared agricultural products, but manufactured foods and home-prepared dishes present even greater difficulties. This knowledge inevitable raises the question of whether we should use these tables for determining the nutrient intake of individuals and groups of individuals. A proper answer to this question depends on the purpose of the study (see Chapter 10) and the information available in the national or regional food composition table or nutrient databases, which comprise the digital equivalent of the table. The information in the following section has to be checked in the local table before starting a dietary survey.

Foods

There will be few problems if listed foods have been adequately described in a comprehensive table. However, in many situations it is likely that there are at least some items missing in a local table. INFOODS (Infoods, 2003) has compiled a worldwide directory of food composition tables which might be helpful in selecting an appropriate alternative source of food composition data. For commonly prepared mixtures or meals, representative data on nutrient composition can be compiled on the basis of recipe ingredients and local cooking procedures.

For specific food items not available in local tables, the choice of an appropriate value should be based on an alternative food. The choice will need to be based as close as possible on a similar species, anatomical part, maturity, age, and method of preparation. However, if the number of foods on which information is not available is very large, methods based on direct analysis should be considered.

Nutrients

The choice of nutrients to be required for inclusion in the table depends on the purpose of the study. For incomplete data sets, borrowing data from other sources may be considered. However, the quality of the data, determined by method of analysis, method of expression (*e.g.* equivalents, international units, or grams), or use of conversion factors (*e.g.* nitrogen into protein), should be checked. When analytical data from more than one table of food composition need to be combined for a single food, such as macronutrients from one table and vitamins from another, it is essential that the data used are standardized to the appropriate level of moisture content, since this clearly affects the level of nutrient content. More detailed information on the quality of food composition tables and their uses can be obtained from Greenfield *et al.* (2003).

It should also be remembered that it is sometimes better to estimate the intake of nutrients indirectly using biological markers and not from food intake. For example, salt (sodium chloride) intake can often be measured more readily by measuring excretion in urine. For the use of bio-markers in nutritional surveys the reader is referred to Margetts and Nelson (1997).

11.3 Conclusion regarding food composition tables

To answer the question of whether tables of food composition should be used for determining the intake of nutrients of individuals and groups, many comparative studies using direct analysis have been conducted, but their conclusions are not unanimous. Marr stated as early as 1971 that comparisons can be made at different levels (Marr, 1971).

Absolute agreement between calculated and measured values cannot be achieved at the individual level. However, differences for most nutrients are almost constant. The least variable nutrient is probably protein and the other proximate constituents. Variation in the vitamin content of foods is much greater and tables are of limited value for minerals. The situation for minerals is complicated by the large variation in the mineral content of foods as well as by the wide variation in the biological availability of minerals in different diets and among different people. However, the entire subject of bio-availability is one that has not been sufficiently addressed.

As yet, in general, calculations using tables and measurements based on direct analyses agree if the food composition table used is compiled mainly from analytical data based on local foods for intakes of groups of individuals. Thus, the food tables used must be matched to foods eaten, especially for staple foods and foods that supply the specific nutrients that are under study. An example of calculations for nutrient intake is given in Appendix E.

12. Meaning and use of anthropometry in field studies on food habits and consumption

Anthropometric measurements are required for a description of the population and a correct interpretation of dietary intake data. Such measurements include height, weight, arm and waist circumferences. Such measurements may reflect the inadequate or excess intake of food, insufficient exercise, and disease. Such data demonstrate that deprivation and excess may coexist across and within regions and even within households. Until relatively recently, attention has been largely focused on infants and young children because of their vulnerability and the value of the non-invasive anthropometry in characterizing growth and well-being. A global analysis demonstrated that child malnutrition is the leading cause of the global burden of disease. This is based on trends in stunting and underweight, amongst others. Stunting is defined as low height for age and underweight as low weight for age based on less than two standard deviations of the NCHS/WHO international reference population (De Onis and Blössner, 2003). WHO recognized, however, the relevance of anthropometry not only in childhood, but throughout life for reflecting the health status and social and economic circumstances of population groups (WHO, 1995; De Onis *et al.*, 2004).

A full appreciation of the utility of anthropometry requires an understanding of the organizational levels of human body composition and the relationship with anthropometric measurements (WHO, 1995). When discussing the measurements, we will explain how different anthropometric measures may indicate body composition.

12.1 Measurements

Anthropometric measurements are not invasive and are simple to conduct. Regular calibration of the instruments and standardization between interviewers are of utmost importance. Small absolute differences in follow-up measurements may have a large health impact and are easily dismissed when observations show random or systematic variation between or within observers. Therefore, a written protocol on the procedure of the technique, including the calibration procedure, should be available for the field survey.

The basic anthropometric measurements considered here are *weight and height*. Stature (standing height) and supine length are considered indicators of general body size and bone length. Assessment of height alone in association with age may be used as an indicator of group nutritional status and will estimate past and chronic malnutrition rather than present nutritional status. The disadvantages of using height alone are that retardation takes time to develop, may be due to genetic differences, or, when found in infants and young children, may be the consequence of small size at birth rather than an indication of postnatal malnutrition.

Height is most cheaply measured under field circumstances by a microtoise or portable measuring rod in sections. It is preferred that the microtoise is fixed to a wall (to the nearest 0.1 cm). The subject should stand (without shoes) on a horizontal platform with his heels together and with the Frankfurter plain horizontal. The subject draws himself to full height without raising the shoulders, with arms and hands hanging relaxed and the feet flat on the ground.

Under the age of 2 years, supine length, or crown-rump length, should be measured to enable comparison with the reference population. There are different types of instruments with a movable foot board or with movable head and foot contacts. Carefully read the instructions included with the instruments to reduce reading errors as much as possible. Height might also be difficult to measure in elderly people due to kyphosis. Total height might be derived from knee height, but this measure might also be biased (WHO, 1995). For measurement procedures see Appendix F.

Body weight, which measures practically the total body mass, cannot by itself provide any information on body composition. When assessing malnutrition, we might be interested in the relative size of muscle, water, or fat mass. A single measurement of weight has to be compared with a reference. Gomez *et al.* (1955) introduced weight for age references for the classification of malnutrition. The simplest way of classifying a child population is by using the percentages of the reference value as class limits, or by using Z scores. The latter can be derived from the mean and several standard deviations of a reference population. The disadvantages of using only weight as a measure for nutritional status is finding a "normal reference population" without under- or overnutrition. Further, oedema may compensate for a weight deficit and the child may be small since birth rather than underweight. Keller *et al.*, (1976) therefore suggested that weight be compared with a reference group of the same height rather than the same age. Weight for height can be expressed as a percentage of the median height at each age. Experience has shown that 80% of the standard is an adequate limit between the malnourished and the adequately nourished. Weighing people seems to be a simple measure. However, this measure is also prone to errors. Thus, an appropriate and repeatedly calibrated weighing machine should be used and bodyweight should be assessed with the people in standard condition. Subjects are usually asked to be clothed in only a light undergarment and to empty their bladder before weighing.

In children, weight and height are used or transformed in indices for the classification of malnutrition. The three indicators preferred internationally are weight for age, height (length) for age, and weight for height (length). The National Center for Health Statistics of the United States of America (NCHS) produced reference sets of data and *growth charts* which are frequently used. If funding permits, a regional set of reference data may be produced which is more in agreement with genetic and environmental features of the target population. A more ambitious goal than just describing "how children grow in a particular region and time" is to develop a reference, indicating how all children should grow when their needs are met. A project has been undertaken for this purpose by WHO in cooperation with the United Nation University (UNU), and is known as the Multicentre Growth Reference Study (MGRS). In addition to including children from around the world in this study, the project also tries to include links between physical growth and motor development to give better insight into the needs of the world's children (De Onis *et al.*, 2004). Recently these standards have been published (WHO, 2006).

For children under five, accurate age is required for deciding whether a child has to be measured standing or reclining, and for converting height and weight into the wasting and stunting indicators. Where there is a general registration of births and where ages are generally known, the recording of age is a straightforward procedure, with age measured to the nearest month or year completed. It is difficult to measure the age of children in many rural areas in developing countries because of the lack of birth certificates and official documents. If this is the case, a local event calendar will need to be designed to show the dates (month and year) of major (local) events in the previous five years. These events can be droughts, floods, festivals, and elections, for example. The month can be determined by matching the birth with seasonal factors such as the dry season, the long rains, the short rains, the maize harvest, *etc.* The actual month and year and, if possible, the day of birth must be recorded, together with the date of measurement. The precise age should be calculated by a data analyst using a computer programme and it is important to never rely on enumerators to calculate the age of the child in the field, as this is very prone to error. If it is impossible to determine age accurately, then it is advisable to select all children with length measurements of 120 centimetres or below. Chances are that most of them will be under five years of age. Only the indicator weight-for-height can be determined under these circumstances, as it does not require age data.

Weight/height ratios are used to provide indications about body "build" or shape and leanness or fatness. The most frequently used ratio is the *Quetelet or Body Mass Index (BMI.)* To obtain the body mass index, weight is divided by the square of height. Several nomograms have been produced to obtain BMI in kg/m^2. Figure 12.1 shows the nomogram developed by Deurenberg. Classification of over and underweight adults is given in Table 12.1. Note that these classifications are mainly based on data from Caucasians, and that other cut off values might be preferred for different ethnic groups

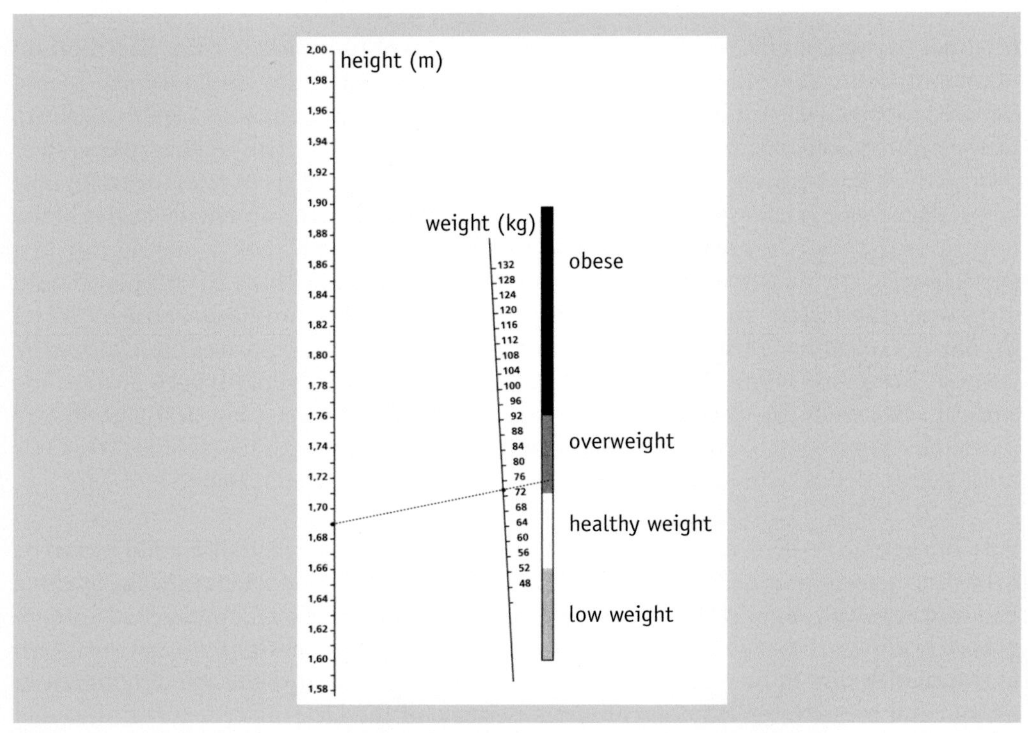

Figure 12.1. Relation between height, weight and overweight and obesity. Source: P. Deurenberg, Wageningen University, Netherlands Heart Foundation.

Table 12.1. Classification of overweight in adults according to body mass index (BMI) (WHO, 1995).

Classification	BMI (kg/m²)	Risk of co-morbidities
Underweight	< 18.5	Low (but risk of other clinical problems increased)
Normal range	18.5 - 24.9	Average
Overweight	>25	
Pre-obese	25 - 29.9	Increased
Obese class I	30.0 - 34.9	Moderate
Obese class II	35.0 - 39.9	Severe
Obese class III	>40.0	Very severe

(Deurenberg-Yap and Deurenberg, 2003). BMI has also been used for older children and adolescents, but not widely for young children because of its variation with age. Many studies have documented that the relative risk of a high BMI may become less pronounced with aging. Several explanations for this phenomenon have been forwarded, such as selective survival, cohort effects, and/or a ceiling effect of mortality rates (Seidell and Visscher, 2000). It might also be that BMI is not a good indicator of body composition in old age. A high BMI in younger adults has a positive relation to fat mass and mortality, while the ratio between lean body mass and fat mass in older adults undergoes change, which is not measured by BMI.

12.2 Other anthropometric indices

Nutrition textbooks provide an extensive description of other anthropometric measurements (Fidanza, 1991; Gibson, 2005). Some will be mentioned here, although few have achieved such widespread use as the height and weight based measurements.

Mid-upper arm circumference (MUAC) has been proposed as an alternative index of nutritional status when the measurement of height and weight is difficult. Under such circumstances, MUAC based on a fixed cut off point (*e.g.* 12.5 cm) has been used as a proxy for low weight-for-height, or wasting. A comparison of the indicators shows a low correlation, however, but MUAC appears to be the superior predictor of childhood mortality in community-based studies. The MUAC can be measured as shown in Figure 12.2.

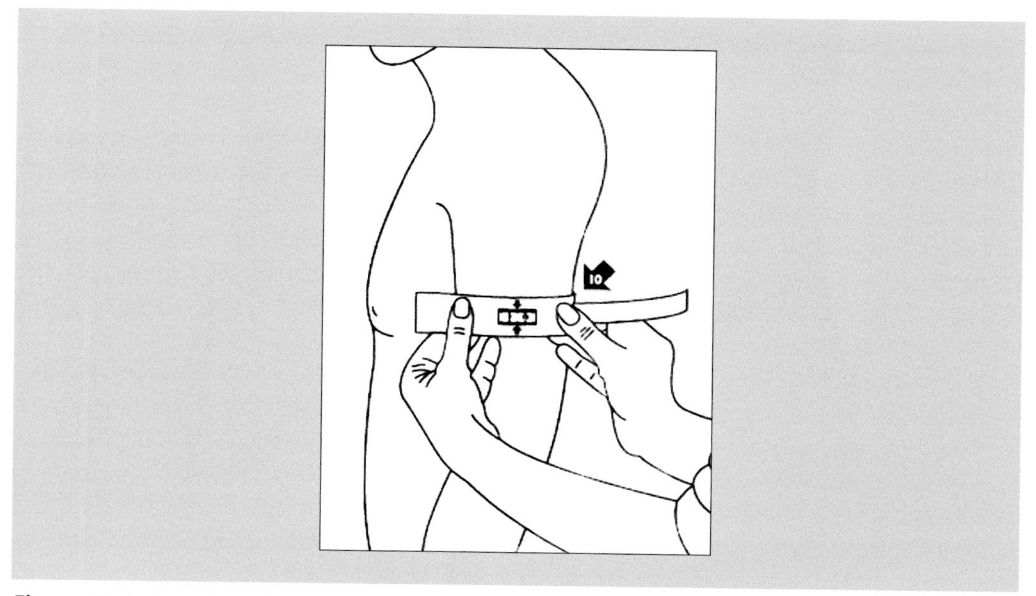

Figure 12.2. Arm circumference measurement.

Arm muscle circumference (AMC) can be derived from MUAC with the help of a triceps skinfold measurement (see Figure 12.5). If we assume that the midarm is a cylinder and covered by a layer of fat, which is measured by the triceps skinfold (TSF), then the AMC is calculated as follows:

AMC = MUAC – (4.18 x TSF)

The combination of the triceps skinfold measurement and MUAC provides a simple field estimate of fat mass and muscle mass (Heymsfield *et al.*, 1982).

Head circumference (occipital-frontal circumference) is often used in clinical settings as part of health screening for potential developmental or neurological disabilities in children. The measurement is of less value for assessing nutritional status in the community.

Waist circumference is a useful indicator of intra-abdominal fat and thus of central obesity. It is a simple measurement, taken mid-way between the lower margin of the last rib and the crest of the ilium in the horizontal plane. The tape should fit snugly, but not so tightly as to compress soft tissues (Figure 12.3). The circumference is measured at the end of normal expiration (WHO, 1995). Suggested cut off values for central obesity are 102 cm for men and 88 for women.

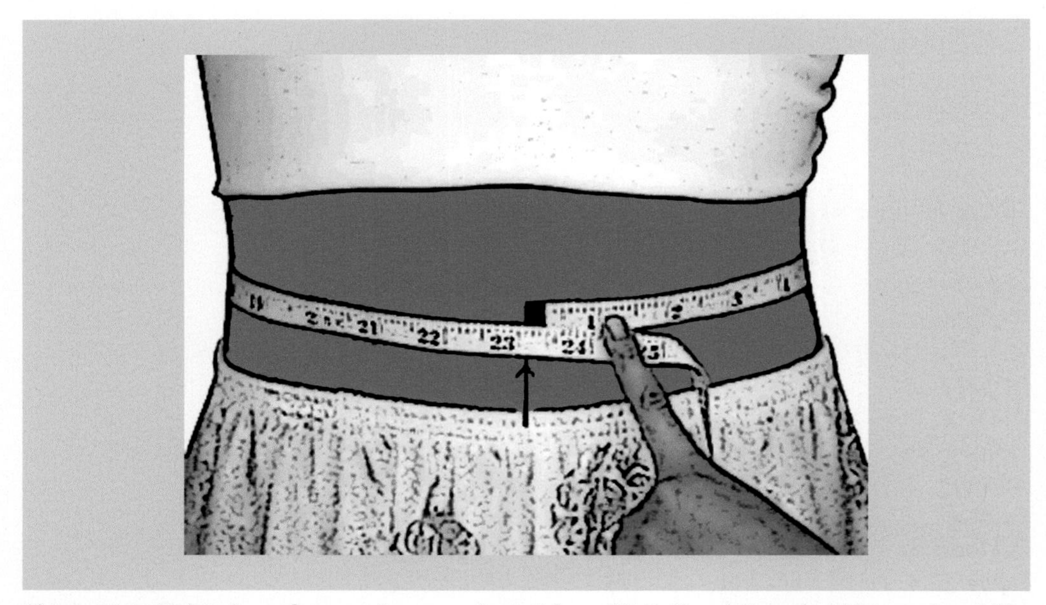

Figure 12.3. Waist circumference. Source: adapted from UN National Household Survey Capability Programme, 1986.

Calf circumference is now considered to be the most sensitive measure of muscle mass in the elderly. It indicates fat free mass with aging and with decreased activity. Calf circumference is taken with an inelastic flexible measurement tape and placed around the calf at the largest circumference.

The *thickness of a fold of skin* picked up at strategic sites indicates the amount of subcutaneous fat, because most of the fat is stored immediately under the skin. Based on the studies of Womersley and Durnin (1977), it is generally accepted that the sites most indicative for fat mass are the following: biceps, triceps, subscapular and suprailiaca. However, in larger field surveys, only triceps, or triceps and biceps, are used to get an impression of the mean fat mass of population groups. The measurement of skinfolds (see Figures 12.4a and 12.4b) is difficult and subject to between-fieldworker error. Therefore, training is required.

12.3 Anthropometric training sessions

Although only easy to handle methods have been selected for this field guide, all measurements are prone to error and may attenuate or even bias results. Inter-observer errors can be large (Fidanza, 1991), and training sessions for fieldworkers are required to

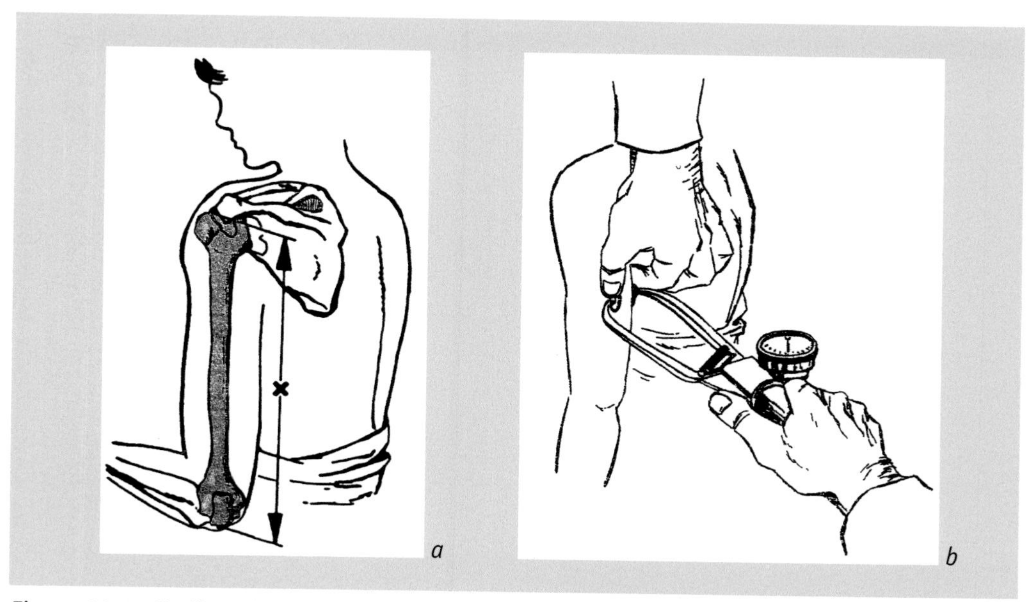

Figure 12.4. Finding the mid-point between the acromion and olecranion for triceps skinfold measurement (a). Raising the fold of skin and subcutaneous fat to measure the thickness with callipers (b). Source: adapted from UN National Household Survey Capability Programme, 1986.

prevent several types of errors. These sessions may explain several aspects of the study, but should include anthropometry, even when fieldworkers have experience from similar studies. The following aspects of measuring should be discussed during anthropometric training:

- the meaning of the measurements for the particular study;
- the measurements to be conducted;
- why standardization between fieldworkers is important;
- what the consequence of within-fieldworkers error can be;
- what the consequences of non-calibrated instruments can be;
- how the measurements should be recorded (see appendix F3)
- the use of Epi-info or other data collection systems; Epi-info is a self-explanatory computerized data collection and analysis system. It is free of charge and available via the WHO site www.cdc.gov/epiinfo/

Procedures for fieldworker standardization in the field collection of anthropometric data is available from the WHO (1995). It is recommended that training sessions be repeated periodically if studies take longer than two months.

In sum, anthropometry is the single most universally applicable, inexpensive, and non-invasive method available to assess nutritional status in the community. However, measurements are prone to error. Therefore, fieldworkers should be trained and provided with written instructions. Appendix F provides an example of a data collection sheet.

13. Some notes on field studies on food habits and food consumption

13.1 Field studies: an outline of steps involved

After having defined the research question, people are often eager to implement the fieldwork. Before planning a survey, any available information concerning the target population must be collected and analysed, including: the geographical context; seasonality; the socio economic and cultural background; and food and nutrition data. Information and data can be obtained from:

- Local resource persons such as local administration officers, agricultural, health, nutrition and social welfare officers, and local leaders who are both women and men.
- Publications and reports, articles in journals. It is also worthwhile to consult websites, both national and international. International organizations and institutions have interesting websites at their disposal, and useful data can often be freely down loaded. See also the list of references.
- The so-called *grey literature*, unpublished reports and documents and statistical data collected by government agencies and NGOs. These hidden documents are often kept in drawers and bookcases and should not be neglected. Because of the frequent turn over of staff, people often do not know that such documents exist, so one has to be very persistent in looking for the material.

It frequently happens that institutions in industrialized countries will have more information on a particular area at hand than what can be found in the concerned developing country.

A food habits and food consumption survey, as well as an anthropometry study, requires several steps to be taken in preparation (just as in any anthropological or geographical survey).

Defining the research question or problem

- What is the precise formulation of the question (research problem) to be answered, or what do you want to know?
- What facts struck you and made you ask the question?
- What is the objective of the survey?
- Who will utilise the results obtained in the survey and in the report?

Survey methods to be used

- What data or facts do you have to collect in order to answer your question (research problem)? This may include the following items:
  Demographic: Date of birth, place of birth, sex.
  Socio-economic: Education, occupation, religion, language.
  Food habits and Consumption: Kind of food, quality of food, food preparation, fuel for cooking, food ideology, food avoidances, food preferences, knowledge on food, food distribution, infant feeding.
- How to collect the data? The choice of methods depends on the source and kind of data or facts one has to collect:

Source		Method
Documents:	studies, reports, files, available statistics	Content analysis Reprocessing of data
Informants:	local leaders, field officers	Interviews
Population		a. Observation b. Non-participant observation on social situations concerning food c. Participant observation, one is temporarily insider by taking part in various activities d. Personal interviews, by means of open questions e. Personal interviews, by means of a structured questionnaire f. Group interviews

Sampling

- *Choice of period of the year*: Agricultural season, market cycle, period of fasting
- Festivities.
- *Sampling of the area*: rural/urban/shanty town, cash crop/mainly food subsistence, other geographical zones (forest, savannah, coastal, lowlands, high lands).
- Sampling of the population.
- *Random sampling*: theoretically equal chances of being chosen, *e.g.* every tenth or fiftieth from a list of households or chosen from a map.
- *Stratified sampling*: samples taken with common characteristics, *e.g.* small farmers, landless farmers, urban workers, educated mothers, the same age and sex.

Organizing the survey, which includes:

- Careful explanation of the objectives of the survey to the interviewers.
- Careful training and instruction of the interviewers on the questionnaires and the way to approach and interview the respondents. The quality and validity of the survey depends highly on the skills of the interviewer.
- Informing the community officials, local leaders, and the population about the coming survey on time.
- Logistical organization; transport of the interviewers and materials.

Actual collection of data in the field, which includes:

- Pilot or trial survey in order to test out the survey method and adjust the questionnaire when necessary.
- Actual data collecting.

Reporting:

- Tabulation and writing up of the data.
- Reporting the outcome of the survey; major findings, policy links, recommendations when required.
- Feedback to the local population on the outcome of the survey, including local authorities, concerned organizations, and high-level authorities.

13.2 Sampling of the population

In selecting households, communities, and their members for food studies, a number of points have to be dealt with. Serious attention should be given to the choice of the period of the year, sampling of the area, and the choice of population. Food habits and consumption are strongly influenced by the time of the year. It makes quite a difference in interpretation if data were collected before or after the harvest (agricultural season), or during periods of feasting or fasting (religious cycle). The market cycle also influences food habits as well. Generally speaking, the agricultural season, or seasonality in food availability, is less pronounced in the urban areas, but it continues to be a point of concern in the urban context. Food or certain foodstuffs may still be available on the urban market, but the price may become so high that these foods are not affordable for the slum dwellers.

Area sampling

Sampling of the area depends on the choice of the research population, *e.g.* cash crop growing communities, areas with mainly food subsistence communities, peri-urban areas, market towns, industrial cities, and the slums. The following items have to be considered in area sampling:

- The geographical zone in which the community is situated, such as the savannah or forest zone (climate and vegetation), coastal regions, lowlands, highlands, or high plateau (relief and terrain).
- The socio-economic status of the community, poor living conditions, availability and access to facilities such as reliable drinking water, health services, schools, *etc.*

After having selected the community and its area and the time of the year for the survey, the research population has to be defined and selected. Usually the selection of an entire population is not possible because of the costs and time involved.

Population sampling

Population sampling is a complicated matter and ideally it should be done in close collaboration with a statistician, as there is always a risk of introducing a bias in the sample. Several methods of sampling exist, but for practical reasons, two sampling methods will be briefly discussed.

Random sampling is a method frequently used in field studies. Depending on the required size of the sample, one can choose as many households as one can handle. It is advisable to draw a simplified map of the community with the houses and then to pick out at random every fifth or tenth household. Houses are numbered in the rural areas of some countries, making a random sampling more easy to do. It is possible to choose from a list with household numbers.

It is necessary to define what a *household* is for the purpose of a food survey. Several concepts and definitions on households can be found in the literature. Households are co-residential units, usually family based and which take care of the resource management and primary needs of its members (Rudie, 1995). Households are not necessarily homogeneous units, having a single set of objectives and preferences. The internal division of labour and economic responsibilities are determined by age, gender, and prevailing culture. To a certain extent, household members are bound to each other by common moral principles which explain the care practices and interests of their members (Pennartz and Niehof, 1999). Households coordinate the preferences, practises, and interests of its members to a large extent (Niehof and Price, 2001). Within the context of a food survey, a household can be defined as a group of persons living under one roof or sharing the same compound and eating from the same cooking pot on a regularly basis. This definition differs from the anthropological definitions of what a household is. From an anthropological point of view, households are defined on basis of family ties and shared economic bonds.

Stratified sampling is another method of sampling and involves choosing a specific group, *e.g.*, a mother and young child, unemployed households, street food vendors, or consumers of street food. Choosing these groups is often only possible with the help of local experts or resource persons who are well acquainted with the situation and the research population.

13.3 Methods of data collecting

In addition to the methods mentioned in chapters 9 and 10, two methods of data collection used in the social sciences are appropriate within the context of a food habits and food consumption survey:
- Observation of the research or target population and its environment.
- Personal interview aided by a questionnaire.

Each method has its weaknesses and strengths, so a combination of both methods can improve the quality of the survey. Surveys based on structured questionnaires should always be complemented with community observation. These observations are very valuable when working out and interpreting the results of collected data.

Food observation

Observational data collection methods are techniques for gathering qualitative data by watching the behaviour of individuals without direct questioning. This method is much used in anthropological surveys and gives useful information on the food habits of a particular community and its members. A disadvantage is the often time-consuming nature of the method. To adequately observe food habits, one needs to stay for some time in the community, whereas food and nutrition programmes often demand quick information for planning and implementation. The presence of an outsider in the community or household will interfere with daily life. People may refrain from eating low esteemed foods for reasons of dignity or vice-versa, making the diet more varied than they can afford on a regular basis. In the long run this will change, as the interviewer will gradually lose the role of guest and become a normal member of the community. Other disadvantages are that only a small number of persons can be observed adequately, and the collected data are difficult to generalize. An important advantage of the observational method is that information can be gathered on items difficult or sometimes impossible to collect by means of questionnaires. For example, respondents will often not fully report on sensitive issues such a bias in food allocation detrimental to women or girls. Items for observation can be found in Appendix A.

Food questionnaires

The food questionnaire allows for the collection of numeric data, which is comparatively quicker to do than the qualitative surveys. Surveys based on questionnaires should not omit observational methods for the reasons mentioned above. The food questionnaire also has several limitations, the most important of which is that not all data can be collected through interviews. Respondents may withhold information for various reasons, or the question may not be fully understood, reasons of self-protection, or possibly the question is considered as too rude or produces social shame.

The construction of a questionnaire is a complicated matter and requires careful attention. Careful attention should be given to the wording of the questionnaire, which should be placed in a logical order. Questions should not suggest an answer. The questions may be pre-coded or include open questions, depending on the nature of the survey. A questionnaire often consists of a combination of pre-coded and open questions. In the pre-coded question, the possible answers are already mentioned and the interviewer needs only tick off one of them. Pre-coded questions can only be set if the researcher already has insight and knowledge of what possible answers may be. In most cases, space has to be given for the category "other", or an open-end question included on the questionnaire.

Several properties should be taken into account when constructing a questionnaire. The instrument used should be valid, meaning that it really measures what you want to measure. It is important to consider whether the questionnaire covers all factors that are relevant to the subjects (face validity), and whether the topics are appropriate, important and sufficient for the setting and the type of subjects (content validity). Criterion validity is also often used as an indicator of the quality of the questionnaire (meaning whether the instrument correlates with a gold standard or superior measure), although this may be less applicable in the social sciences. Furthermore, the questionnaire should be appropriate and acceptable. This includes not being too short or too long, and that the format and the questions should be acceptable and suitable. It is also important to look for existing questionnaires and to determine whether they work, although copying or translating existing questionnaires does not guarantee the usefulness of the questionnaire in your study setting. The questionnaire developed should also be reliable and should produce the same results when repeated in the same population.

There is always a tendency to introduce all kinds of items when constructing a questionnaire. Every item should only be considered for inclusion if it relates to the research question. In other words, the researcher should ask if the item will provide an answer to what is really needed to be known. Every additional question increases the time needed for interviews and makes people less willing to cooperate. It also increases the time needed for tabulation and analysis. Questionnaires should be limited in length and scope to the required essential information. Examples of a food questionnaire are presented in Appendix B. Prudence is called for as the food questionnaire is an example and can only be used when adapted and changed according to the specific research questions and objectives. Appendix B starts with demographic and socio-economic questionnaires (B1). The other sections deal with food supply and food preparation, distribution of food, infant feeding, food avoidances, and special foods (B2-B7). The questionnaire concludes with a section on the frequency of food consumption (B8). These types of data provide information on the kind of food consumed and give a picture of the dietary pattern of the research population. It is advisable to collect food frequency data from a selected number of the research population after completing the previous sections of the questionnaire.

In most cases, a survey has to be executed in another language, the language of the respondents. In many cases it is difficult to translate the questionnaire into another language, as certain concepts cannot be translated literally. In many instances the concept "food" differs greatly in many cultures. In Java, for example, there are different names for the staple food rice, including rice as a crop, rice as a seed, and rice as a food. In several African languages the name food is a synonym for "staple food" and not necessarily for vegetables or condiments. Questionnaires should be translated with the help of persons familiar with the local situation, language, and idiom. It is advisable to keep a detailed diary or logbook during the survey to record observations made on food and nutrition, and on information received from various informants and the population. The diary can be very helpful for the interpretation of the survey results and in drawing the final conclusions.

13.4 Fieldwork

Before starting with the implementation of the survey, the questionnaire's feasibility needs to be tested in a pilot survey and modified when necessary. The pilot survey also serves to train the interviewers. The quality of the data depends heavily on the skills and conduct of the interviewer during the survey, rather than on the respondents. So the selection and training of suitable interviewers is crucial for the survey. Is the interviewer likely to be accepted by the research population and able to handle confidential information? Even a semblance of being talkative or showing careless behaviour with information about other households will jeopardize the survey. Gender is of great importance in the selection of interviewers. Food and nutrition is a woman's responsibility in most cultures. Male interviewers are often not acceptable for making interviews at household level in traditional communities, so most interviewers are likely to be women. Interviewers should be aware of the dynamics of an interview, and they can practise this by first interviewing each other. By interviewing each other, one becomes more aware of the aims and method of the survey, how respondents should be approached, how possible embarrassments about poor diets can be minimized, and the importance of maintaining confidentiality during the interview.

The *code of conduct* during the survey implies that the interviewer should:
- Be fully familiar with the purpose and significance of the survey.
- Have a good rapport with the respondent and the community.
- Show a large measure of respect for other people regardless of social background, position, religion, and ethnicity.
- Begin the survey with polite greetings and friendly conversation, but not talking too much.
- Explain the purpose and the content of the questionnaire, and emphasize the importance of the answers given.
- Make records of striking observations.
- Treat the collected information confidentially.
- Thank the respondents before leaving.

Interviewers should make note during the survey of the selected persons or households who refuse to take part. Reasons for refusal have to be recorded. Such information can be useful in finding out whether the sample was biased in spite of all efforts to obtain a random sample. A number of obstacles are commonly encountered in collecting field data. Local people are often suspicious of strangers and may therefore be unwilling to communicate fully with the interviewer. Such attitudes may be caused by previous unpleasant experiences. Respondents may finally talk to the interviewer in such a situation, but they try to protect themselves by deliberately giving wrong information or answering rapidly without thinking to get rid of the interviewer as soon as possible. Of a positive nature is the habit in many cultures to please the interviewer by giving the expected answers instead of reporting on their factual food habits. People are often not used to responding to a whole range of questions and may retreat from the situation by stating, "I don't know", or "I can't remember". When the fieldwork is finished, the respondents expect to receive a token of friendship from the researcher and interviewers in the form of a present. It is a kind of compensation for all the effort given by the respondents to provide the survey team with the needed data and other kinds of information. In most cases, it is sensible to give a non-food present during (when required) and after the survey. Food may create a bias in the research findings.

Each questionnaire should be completed on the day of collection and be carefully checked by the interviewer and the supervisor to detect any inconsistencies, errors, or incompleteness. At this stage of the survey it is still possible to visit the respondent again and to check statements and to make the required corrections. After the fieldwork is finished, it is too late to do so. For practical reasons the supervisor should place his signature on the questionnaire, indicating it has been carefully checked.

13.5 The backpack nutrition library

For relatively short-term field studies in areas where electricity and internet connections are not near at hand or are unreliable or unavailable, it can be useful to bring along lightweight and carefully selected reference materials such as:

- *Food habits and consumption in developing countries*
  Manual for field studies
  Wageningen, Wageningen Academic Publishers, 2006.
- *An appropriate food composition table*
  Food composition tables are listed at the FAO website and some can be downloaded; www.fao.org/infoods/directory_en.stm
- *Nutritional requirements*
  Depending on the nature of the field studies, appropriate dietary requirements can be downloaded and printed from the FAO database. For human energy requirements: www.fao.org/documents/show_cdr.asp?url_file=/docrep/007/y5686e/y5686e00.htm;
  For micronutrients and mineral requirements: www.fao.org/documents/show_cdr.asp?url_file=/docrep/004/y2809e/y2809e00.htm;
  For protein requirements: www.fao.org/docrep/003/aa040e/aa040e00.htm
- *National food balance sheet*
  Food balance sheets showing food and nutrient availability of a country can be downloaded for most countries from the FAO database. For conducting field studies, a copy of the food balance sheet may serve as a reference point. Website address: faostat.fao.org/faostat/form?collection=FBS&Domain=FBS&servlet=1&hasbulk=0&version=ext&language=EN

14. Reporting and dissemination of survey findings

When data have been collected, they must be tabulated and analysed. Tabulation is the summarization in the form of quantitative tables. Almost all the information collected by questionnaire can be presented quantitatively. Quantitative analysis of social situations is usually more convincing for policy makers than a descriptive or qualitative approach.

The oldest method of data tabulation is by hand, straight from the questionnaire, by ticking each entry from the questionnaire off on a list. This is simple and useful, not requiring special equipment apart from a scientific pocket calculator. If one needs to work with more than two variables it is quicker and easier to use other methods such as a statistical software programme on a personal computer. One must then discuss the questionnaire with a local data processing specialist before the fieldwork begins. However, hand methods may suffice when collecting basic information on the food habits of a few persons.

14.1 Presentation of numeric data on food habits

The findings of the survey may be presented in tables, graphs, maps, sketches and photographs. Most findings may be presented in tables, but graphic presentation may make the data easier to understand.

A *table* has three main parts: the caption (including number, title, and explanation), headings, and the columns with the figures themselves. A *bar* or a *line* chart enables you to visualize a comparison, such as between years/months or gender, or between a target and the actual situation. These types of charts can also be used to show a declining or increasing trend of the variables studied. A *pie chart* can be used to show disaggregation, proportions, or contrast (rather than trends). A *map* is a flat representation of what happens on the ground. A map visualizes selected information related to the geographic location. It helps you to visualize a situation quickly, it facilitates the analysis, and it helps planners to see where resources/ attention is most needed. With a map, you can show the location of communities, an attribute value to communities, the location of basic infrastructure, the radius of access to infrastructures, a relation between two or more indicators, and a comparison of an indicator between (sub) districts or regions.

Every table and graphic presentation or figure should have a title, placed above the table or figure, and should define the content briefly and clearly. Explanatory detail may immediately follow the title or may be placed at the bottom of the table or figure. If the information is obtained from other studies, the source should be mentioned at the end of the caption. Each table and figure must be understandable independently of the text of the report. Much of the data on food habits can be presented either descriptively or quantitatively. The two forms of information should complement and not duplicate one another. The advantages of tables and figures are that they reduce explanatory and descriptive statements, facilitate comparison, and make it easier to remember data. Descriptive and quantitative data should complement each other for a good understanding of a food and nutrition situation.

14.2 Working out trends

If baseline data is available, one can work out changes in food habits within an intervention period. To allow valid interpretation, one must compare earlier data with recent data collected by the same method in the same target group, and compare them with data from a reference group (Figure 14.1).

A reference group should be as identical as possible to the target group, but differ mainly in that it has not been covered by food and nutrition activities. A comparison of A1 with A2 tells us whether any changes in food habits have occurred over the intervention period. A number of questions arise from this: Is the change the result of a general improvement in the economic situation of the district, region, or nation? A result of favourable weather? Or a result of food and nutrition activities? It is necessary to look carefully into these questions.

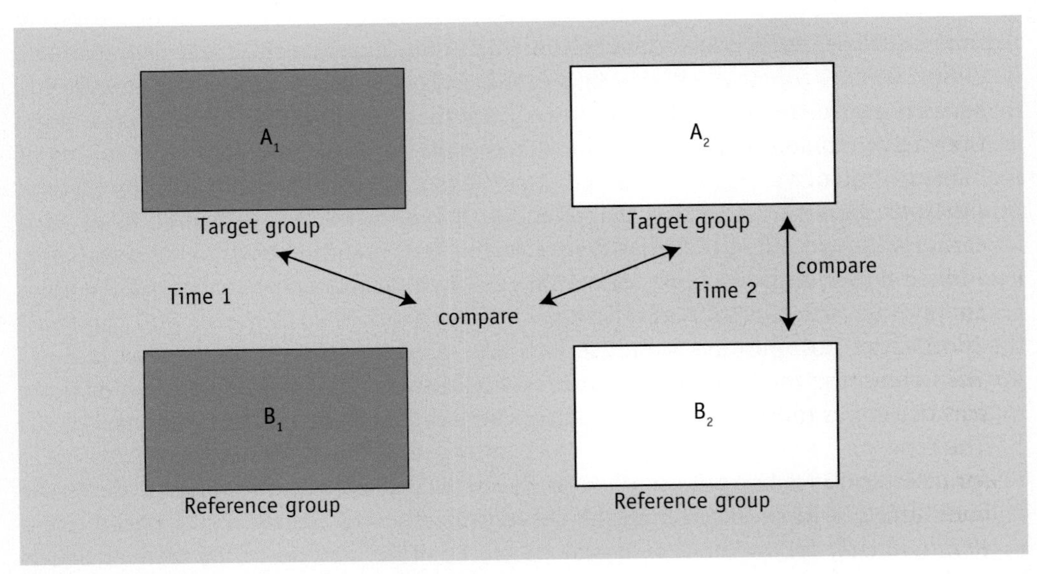

Figure 14.1. Measuring change in community A.

To clarify the observed change, one must use a reference group. Under ideal circumstances, baseline data (B1) should have been collected from the reference group in order to detect change. Let us assume that the reference group B2 differs significantly from A2, but resembles both the baseline situation of A1 and B1. The change in food habits may then have resulted from food and nutrition activities in the district or region, rather than from a general economic improvement. If there has been a general economic improvement, the reference group is likely to change in the same direction as A2.

Unfortunately, such baseline data have not been collected in many cases. In order to get at least some idea of possible changes in food habits within an intervention period, one may compare A2 with the reference group B2. However, working out trends and detecting intervening variables is a very complicated matter so that collaboration with a statistician is recommended.

When no baseline data are available, it is also possible to measure a trend in a qualitative way by means of the previously discussed method of a three-generation study (see section 9.3). This approach will give an answer to the question of whether a specific nutrition problem is only a recent phenomenon, of a chronic nature, or the outcome of a long-lasting downward trend.

14.3 Putting the report together

There are several ways of preparing a field study report. The following successive elements are proposed for a report on a single field study on food habits and consumption along the lines we have discussed:

- *Abstract*: All the bibliographic and indexing information needed by secondary services grouped around an informative summary. Policy makers and administrators may need to read your summary, so do not make it too technical.
- Table of contents.
- *Introduction*: General circumstances and reason for the study, and reference to any earlier studies that have been published.
- *Aim or objectives*: Reasons why the study was carried out and its purpose.
- *Survey methods*: Choice of methods for the survey including time and place as well as limitations of the survey.
- *The community and its situation*: The food habits and consumption of a community or part of a community, such as men, women or children, cannot be studied in isolation. The report should briefly describe and discuss the social and physical environment, for instance the ecological zone, the socio-economic situation, and the relation of the community to the outside world.
- *Results and discussion*: Main findings compiled around tables of data with discussions of their significance.

- Conclusions and (if appropriate) recommendations.
- *Acknowledgements*: A courtesy and a token of gratitude, acknowledging assistance that contributed to the study.
- *References*: A reference should provide the elements needed to identify and obtain any written work cited in a clear and succinct form.
- *Appendix*: Bulky material not forming part of the main line of argument may be placed at the end, for instance a glossary of foods and the questionnaire used.

14.4 Dissemination and use of results

One of the main challenges of a study is to keep from being shelved without being used. Three determinants are important to ensure that your study is going beyond data collection. First, the ownership of your study is important. The more the sense of ownership, the higher the chance that your results will be used. To increase ownership, it is important to involve all stakeholders already in the development phase of the study. Stakeholders include both beneficiaries as well as programme implementers. The relationship with the stakeholders should be maintained throughout through dialogue and interim presentations on the process and preliminary outcomes of the study.

A second important determinant is the usability of the results. The credibility of results plays a large role in their usability and can be improved by using reputable information sources, assuring that the information is accurate and unbiased. The findings should also be informally discussed with stakeholders before finalizing the report, and their comments and suggestions should be included. Results will also be more usable when presented in time. Results should be available at the time when decisions are to be made and there should be a short time interval between collection and dissemination.

A third determinant is the effective presentation of the results. Results can be disseminated in a scientific report, but this will be read by a limited number of people. It is therefore important to disseminate your results in different ways tailored to the different stakeholders of your study. The dissemination can take place at different levels, such as at the community level where results are discussed in a community forum, or at the smallest administrative level in a small workshop or seminar. Posters or short leaflets may help to effectively disseminate your data to keep your report from being shelved.

PART THREE:

APPENDICES

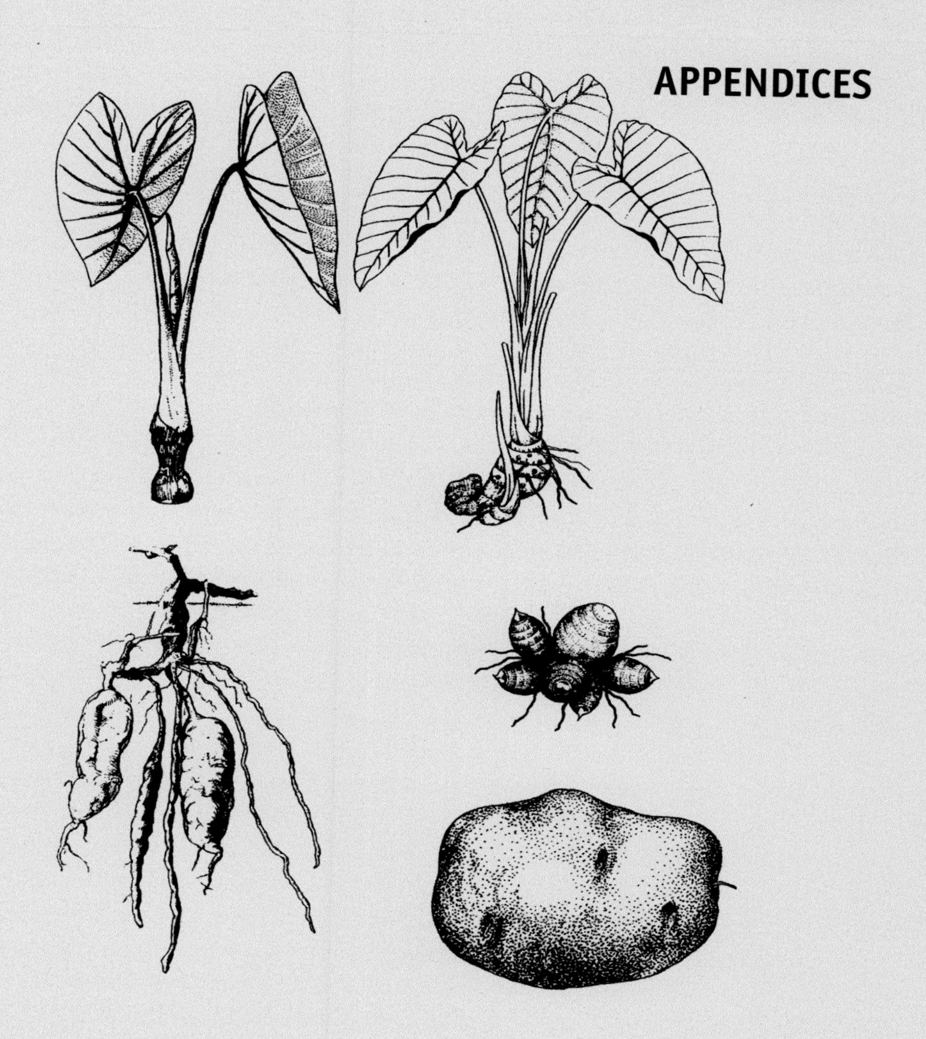

Appendix A:
Items of observational data on food ethnography

This list of items on food ethnography is a tool for collecting socio-economic and cultural data on food and nutrition by means of observation. It is advisable to use the list after having defined the objectives of the survey. When observing the community and the survey population, it is important to give adequate attention to the gender dimensions of food, nutrition, and related matters. This list can also be used as a source for constructing food and nutrition questionnaires.

A1 Geographical context

1. Type of community (village, suburb, quarter of a town, shanty town).
2. What are the means of existence of the community (subsistence farming, cash cropping, fishing, industry and trade in the formal and informal sector)?
3. Means of communication (main road, railway, waterway, airport).
4. Distance from nearest means of communication.
5. Vegetation zone.
6. Geographic zone of the settlement and agricultural fields (lowland, highland, mountain).

A2 Social structure of the community

1. Social status of the survey group in the community (elite, middle class, lower class, tenant, landless labourer, estate worker, caste, ethnic and religious group).
2. Availability and access to social services (schools; medical services including hospital, medical doctor, public health nurse, or dispensary; social welfare programmes, credit facilities, rural extension, nutrition-related intervention programmes).
3. Food marketing services (market, eating houses, ready-to-eat foods and beverages sold by vendors and hawkers (street foods), shops selling food, butcher, baker).
4. To what extent and how does the group use the services?
5. Are "traditional" medical practitioners and other traditional advisers consulted on matters such as health?
6. Who owns (social class, gender) the land, crops, livestock, fishing boats, workshops, industry, houses, and other resources?
7. Who decides on the political and economic affairs of the community inside and/or outside the community?
8. Who are considered by the target population to be their leaders (both male and female)?
9. Are agricultural monetary resources in the community available so that they cover the food needs (in the sense of energy) of all its members? If not, does this apply for the whole community or only certain categories or groups within it?

A3 Food security

A3.1 Food supply

1. What is the agricultural production system and land ownership (Socio-economic class, gender: subsistence farming, cash-crop farming, size of the farm, smallholdings, estates)?
2. What kind of agricultural implements and techniques are used (digging stick, spade, pick, hoe, plough; irrigation, terrace cultivation, crop rotation with fallow)?
3. Which kind of agricultural and related tasks are carried out by men and which by women?
4. What is the agricultural calendar?
5. What kind of crops are cultivated (staples, grain legumes, vegetables, non-food crops)?
6. What kind of fruit-trees are grown?
7. What kind of livestock is kept?
8. What kind of fish is caught?
9. Method of fishing (netting, trapping, poisoning, angling)?
10. What kind of foods (animals and plants) are collected (food-gathering)?
11. What kind of animals are hunted or trapped for food and how?
12. What kinds of food are obtained from the individual's own farm and from barter?
13. Who goes to buy food from the market, shops and other places?
14. What kind of foods and non-food items are purchased on the market?
15. Is food adulteration common practice and, if so, on which foods and how?
16. Who provides money for food on the market?
17. What kind of foods are given away or received as a gift?
18. Are foods exchanged and if so, for what?

A3.2 means and methods of food storage and preservation

1. Kinds of food stored.
2. Where is the food stored (inside or outside the dwelling, in silos, pots or other means)?
3. Methods of preservation, *i.e.* smoking, pickling, fermenting, salting, drying (air, sun, mechanical).
4. Who owns the stored foods and who is in charge of the distribution?

A3.3 Coping with food shortages

1. What kind of foods were consumed during the last period of food shortage?
2. What was the nature of the food shortage?
    a. seasonal, such as a pre-harvest shortage;
    b. non-seasonal, but occurring from time-to-time such as bad harvest;
    c. non-seasonal, but very rare;
    d. other.
3. What kind of foods, if any, are consumed during famine?
4. Are there any particular foods only consumed during famine?
5. Are unusual substances such as bark or clay used as food during famine?

6. Who supported the household during famine?
7. Are the number of meals a day reduced during famine?

A4 Food preparation

1. Who fetches the fuel for cooking?
2. Who fetches the water for cooking?
3. Kind of water supply (piped water in the house, piped water outside the house, river, well and type, other; distance).
4. Is sufficient water available the whole year round?
5. What method is used to make a fire (matches, wood, friction, percussion, hammering)?
6. What kind of cooking stoves are owned and used (electric stove; electric stove with oven; gas stove; gas stove with oven; butane gas stove; butane gas stove with oven; kerosene stove; smokeless stove with oven; charcoal pot; tripod of stones or mud; separate oven made from mud; separate oven made from tin plate; hot box?
7. What kind of kitchen utensils are available and what are they made of (pot, bowl, frying pan, cooking pan, dish, plate, cup, saucer, spoon, fork, knife, ladle, grater, pestle and mortar for pounding grains or roots and tubers, small pestle and mortar of stone or wood, grinding stone: iron, clay, metal, stone, wood)?
8. What type of fuel is preferred and what type is actually used for cooking (electricity, gas, butane gas, kerosene, coal, wood, split wood, branches, twigs, charcoal, straw, dried cow-dung)?
9. How is fuel obtained (collection, purchase)?
10. What is the distance to purchase and/or collection place?
11. Is there a fuel wood shortage for the preparation of food?
12. Who does the cooking?
13. Who assists in preparation of food (daughters, wives, relatives, neighbours)?
14. Where is the food prepared (inside the dwelling, in the open, both inside and outside)?
15. How are the different kinds of food prepared (boiling, parboiling, stewing, steaming, baking, frying, grilling, roasting, smoking, salting, pickling, fermenting, rotting, drying in air, sun or machine, eaten raw)?
16. How are different drinks prepared (solution such as sugar or honey in water, suspension such as meal and water, extraction or treating with cold water, infusion with hot water such as tea, boiling, fermentation)?
17. Is edible earth customary eaten?
18. What kind of salts are used for food preparation (traditional salt *e.g.* sea-salt, inland rock and coarse salt, extract of plant ashes, commercially refined salt, regular or iodized)?
19. What kind of stimulants are used to increase vital and intellectual activities (coffee, tea, cocoa, kola nuts, mate, quinine bark, betel or areca nut)?
20. What kind of narcotics are used to produce lethargy or stupor and relief of pain (opium, coca leaves, qât leaves, Indian hemp)?
21. What kind of digestives are used?

A5 Street foods

1. Are street foods (and other already prepared foods) consumed on a regular basis?
2. What kind of street foods are consumed, snacks, components of a meal (such as the staple only, the meat or fish), a full meal?
3. When and how are the street foods consumed, as breakfast, midday meal, or evening meal?
4. Where are the street foods consumed, at home, in the residential area, at work (or school)?
5. In case street foods are consumed at home, are the street foods used as a meal, as an ingredient (*e.g.* fish, meat), as a staple (*e.g.* rice, cereal porridge) or as a staple with home-prepared components?
6. How do people perceive street foods as compared to home-prepared food?

A6 Distribution of food at household level

1. How many meals a day are served, at what time, or after which kind of work?
2. What are the different eating groups during the meal time?
3. Who is responsible for distribution of food within the household and in each eating group?
4. Does each member of the household eat from his own plate?
5. Do the members of the household eat from a common plate or pot?
6. Is it the habit to wash hands before eating?
7. Is food eaten with the hands?
8. If not, what eating instruments are used (fork, spoon knife, chopsticks)?

A7 Feeding of infants (up to three years of age)

1. How long is breast-feeding continued?
2. Is breast-feeding given on demand of the child, according to a schedule, or according to maternal inclination?
3. Do women have sexual intercourse during lactation and, if sexual intercourse is avoided, why?
4. At what age is the first food other than breast milk given, and what is the method of feeding (cup, cup and spoon, feeding bottle, hand, chewing by mother)?
5. Is breast-feeding increasingly replaced by milk products such as: powdered milk, evaporated milk, sweetened condensed milk? If so, why?
6. Do present duties and responsibilities allow mothers to give sufficient time to breast-feeding and other forms of infant care?
7. Are promotional activities for infant foods going on in the community, and if so, which kind?
8. Why did the mother stop breast-feeding?
9. Was breast-feeding stopped at once or gradually?
10. How is breast-feeding stopped?
11. What kind of special foods for infants (weaning) are prepared or bought?
12. Is forced feeding of infants practiced, if so how and why?
13. Who takes care of the child, especially feeding (the mother or guardian)?

A8 Food avoidances and fasting

1. What kind of foods and drinks may not be consumed by the following categories: infants during weaning; girls, boys, women during menstruation; women during pregnancy; women during lactation; all women; all men?
2. For what reasons are these foods and drinks avoided?
3. What kind of foods and drinks cannot be consumed during illness by adults, by children?
4. What kind of foods may not be consumed during religious feasts, or other special occasions?
5. Are there periods of fasting in a year and is this a community event or done on an individual basis?
6. Does fasting involve all members of the community, or only specific categories such as men, women, initiation groups or priests?
7. Is there in the community a tendency or trend in lessening or abandoning certain rules of food avoidances and fasting?

A9 Special foods and drinks

1. What kinds of special food and drink are consumed by pregnant women?
2. What kinds of special food and drink are consumed by lactating women?
3. What kind of foods and drinks are prepared, purchased or given for: birth of a child; puberty rites of boys; puberty rites of girls; wedding; funeral; sowing or planting ceremonies; harvest ceremonies; other occasions or feast days?
4. What kind of foods and drinks are offered to (distinguished) guests?

A10 What do people think about foods and drinks

1. What kind of foods and drinks do women consider the best and what kind do they consider less good for husbands, wives, pregnant women, lactating women, breastfed children (supplementary foods), infants during weaning, pre-school children, sick children, sick adults, old people?
2. What kind of plants, animals or parts of animals are considered unfit for consumption and why?
3. Is there a system to how people classify their own foods?
4. What is the appreciation for modern processed and packaged (industrial) food and drinks (in bottles, cans, cartons) compared to traditional ones?

A11 Food preferences

1. What kinds of food and drink do men and women most like to eat?
2. For what reasons are these foods and drinks preferred?
3. What kinds of food and drink do men and women not like to eat?
4. For what reasons are these foods and drinks not liked?

5. What kinds of commodity, both food and non-food, would men and women like to buy when sufficient money is available?

A12 Nutritional perceptions

1. Do men/women recognize signs and symptoms of certain nutrient deficiencies such as vitamin A or iodine deficiency?
2. How are these signs and symptoms locally named?
3. What are common beliefs and explanations concerning these signs and symptoms?
4. Are people worried about these signs and symptoms?
5. What kind of actions or measures are taken to reduce the signs and symptoms?

Appendix B:

Examples of food ethnography questionnaires

B1 Household members

This sheet deals with the composition of a household and the questions that should be asked to the head of the household. Within the context of a food and nutrition survey the household is usually defined as a group of persons eating regularly from the same cooking pot. The questionnaire is divided in two subsections: the first one (B1.1) will give information on the demographic background and the second one (B1.2) on the socio-economic background of the household and its members. These are possible variables which may influence the food habits. At the top of the sheet fill in the name of the place, address of the household or a short indication of the situation of the house. In some countries each house has a number. After the questionnaire has been completed it should be carefully checked by the supervisor and when ok, provided with his signature. Do not forget to note down the date of the interview. After the survey is completed, all names on the questionnaires have to be deleted.

B1.1 Demography (see also Form B1.1)

Person number. Each member of the household will be given a *code number* to be used throughout the questionnaire.

Name of each member of the household.

Relationship to head of household. Place each person on a separate line and list the persons in the following order:

1. Male head;
2. Female head;
3. Unmarried children from the oldest to the youngest (mention the relation to the head of the household, son or daughter);
4. Other family members. Start first with married children, their spouses and children;
5. Other members of the household such as employees or boarders.

Sex. Male or female.

Date of birth: age. If known, state date of birth and calculate age in completed years. Age often has to be estimated, which may be quite a complicated matter. National census reports often give information on how to estimate the age of persons who cannot give accurate information on their date of birth.

Marital status. Indicate if applicable, 1st wife, 2nd wife, married, unmarried, widow or widower, divorced.

Residence. Indicate whether permanent or visiting.

B1.2 Socio-economic information (see also Form B1.2)

Person number, name. Repeat as on Form B1.1.
Ethnic group. If applicable, indicate which ethnic group each person says they belong to.
Religion. The religion to which a person says he belongs.
Education level. Final level at primary or secondary school, or other educational institution. If a person did not receive any formal education but can read and write, state as literate and the language in which he or she is literate.
Main occupation. When more than half the total working hours are devoted to one occupation. If no occupation, place a dash (-).
Other occupation. When less than half the total working time is devoted to another occupation. Sometimes the information received may be rather confusing, *e.g.* a person may state that he is a driver, while in fact, he is a farmer and driving is only a subsidiary occupation. Careful questioning is needed to distinguish the main occupation from other occupations.
Employment status. This may be stated as self-employed, government-employed, employed by a private firm in the formal or informal sector, or unemployed.
Place of work. This may be in the community itself or elsewhere.

Form B1 Householdmembers

Place:.....   Address:.....   Name of field-worker:.....   Date:..........

Signature supervisor. ....

B1.1 Demography (information from head of household)

Identification no. of household:.....

Name of head of household: ....

Person no. [Code number]	**Name and residence**	**Relationship to head of household**	**Sex**	**Date of birth**	**Age (years)**	**Marital status**
1						
2						
3						
4						
5						
6						
7						
8						
9						
10						

Form B1.2 Socio-economic information (information from each person)

Identification no. of household:.....

Name of head of household:.....

Person no. [Code number]	**Name**	**Ethnic group**	**Religion**	**Educational level**	**Main occupation**	**Other occupation**	**Employment**	**Place of work**
1								
2								
3								
4								
5								
6								
7								
8								
9								
10								

B2 Food availability

(Information at community level)

The questionnaires B2.1, B2.2-2.3 concern collecting information on which kinds of foods are available at the community level and in which period of the year. These data can be collected by means of interviews of a number of key informants and by observation. It is essential to observe with a key informant the available foods in the market, shops and the crops and livestock in the fields. For this type of information at community level it is not necessary to interview all households, but to limit interviews to a number of key informants, persons who are knowledgeable in this field.

Questionnaire B2.1 is for the construction of a list of foods available in the community. A difficulty encountered is the determination of different kind of foods with its local, English and scientific names. Advice from local experts is essential. Efforts should be made in getting hold on reports and books with information on the names of local food plants. Questionnaire B2.2 deals with the yearly agricultural cycle or calendar of the main cultivated crops of the community. When appropriate, a questionnaire can be made on the gathering and hunting activities. Write down for each main crop the English name and the corresponding periods in month(s) of sowing or planting, weeding and harvesting. Careful attention should be given to the seasonality of food availability and the lean season of the year when food stocks are getting depleted.

Questionnaire B2.1 List of foods

Place:............ Address:.............. Name fieldworker:................. Date:.........
Signature of supervisor:.......

English name	Local name	Scientific name
cereals		
....	....	....
....	....	....
....	....	....
tubers		
....	....	....
....	....	....
....	....	....
grain legumes		
....	....	....
....	....	....
....	....	....
oilseeds and nuts		
....	....	....
....	....	....
....	....	....
vegetables		
....	....	....
....	....	....
....	....	....
fruits		
....	....	....
....	....	....
fats		
....	....	....
....	....	....
condiments		
....	....	....
....	....	....
animal foods		
....	....	....
....	....	....

NB: If focused on a specific nutrient for example vitamin A rich foods, the table may be complemented with the vitamin A content based on appropriate food tables. See further Annex C 21 consumption and attributes of key food items.

Questionnaire B2.2 Agricultural calendar (in months)

Main crop	**Period in month(s)**		
English name	**Sowing/planting**	**Weeding**	**Harvest**
....	....	....	....
....	....	....	....
....	....	....	....
....	....	....	....
....	....	....	....
....	....	....	....
....	....	....	....
....	....	....	....
....	....	....	....
....	....	....	....
....	....	....	....
....	....	....	....
....	....	....	....
....	....	....	....
....	....	....	....
....	....	....	....
....	....	....	....
....	....	....	....
....	....	....	....
....	....	....	....
....	....	....	....
....	....	....	....

Questionnaire B2.3 Periods of seasonal food shortages in month(s):

....

....

B3 Supply and preparation of food

(Information from women responsible for food preparation)

Often these questions may be asked from the wife of the head of the household. If more than one woman is responsible, for instance a second wife, a second sheet of the questionnaire should be used. Bear in mind that assisting in food preparation is not the same as being responsible for the preparation of food. Questions 3.1 to 3.7 may be answered by the family relationship of the person doing each task to the respondent.

Place:.......... Address:........ Name fieldworker:................. Date:..............

Signature of supervisor: ....

Identification no. of household: ....
Name of head of household: ....

Name and person number of respondent	
3.1 Who fetches food from the farm?	
3.2 Who buys food from the market?	
3.3 Who receives food by barter?	
3.4 From whom is the barter received?	
3.5 Who provides the money to buy food?	
3.6 Who fetches the fuel for cooking?	
3.7 Who fetches the water for cooking?	
3.8 Who helps you in preparing the food?	
• daughter(s)	[ ]
• co-wife(ves)	[ ]
• relative(s)	[ ]
• neighbour(s)	[ ]
• others (stated)...........	[ ]
3.9 Is food prepared?	
• inside the dwelling	[ ]
• in the open	[ ]
• both inside and outside	[ ]

B4 Distribution of food

(Information from person(s) responsible for food preparation)

If more than one person is responsible for food distribution, a second sheet should be used. Question 4.1 is concerned with the eating groups during mealtime. It is necessary to know who is responsible for the distribution in each eating group in order to understand the distribution system (Question 4.2). Of equal importance is knowing the order in which food is received (Question 4.3) by the household members and whether the food is eaten from a common dish or each from his own plate (Question 4.4).

Place:.......... Address:........ Name fieldworker:.................. Date:...............

Signature of supervisor: ....

Identification no. of household: ....
Name of head of household: ....

Name and person number of the respondent	
4.1 With whom did the members of the household eat yesterday?	
4.2 Who is responsible for general distribution of food during meal time in each eating group?	
4.3 In what order is the food received by the members of the household? (indicate 1, 2, 3) • husband • wife • son(s) • daughter(s) • others (stated)	 [ ] [ ] [ ] [ ] [ ]
4.4 Do members of the household eat from • the same dish or plate? • or own plate?	 [ ] [ ]
4.5 Is it usual to wash hands before eating?	[ ]

B5 Infant feeding

(Information from mothers of children below 3 years of age, except in some societies where the guardian must be interviewed too)

In some societies, children may be placed in the care of a relative rather than with the mother. If so, the person responsible for the child should be identified and also interviewed if necessary. If there are more children below this age in the household, a separate sheet should be completed for each, with the name and code number of the respondent, and the age and code number of the child.

Place:.......... Address:........ Name fieldworker:.................. Date:...............

Signature of supervisor: ....

Identification no. of household: ....
Name of head of household: ....

Name and person number of the respondent	
Name, age and person number of the child of the respondent (below 3 years of age)	
5.1 Do you breast-feed your child now? (if not, go to question 5.7 and further on)	
5.2 If so, when is breast-feeding given: • On demand of the child? • According to schedule? • According to your inclination?	 [ ] [ ] [ ]
5.3 Do you give your child supplementary food as well as breast milk?	
5.4 If so, at which age did the child receive food other than breast milk?	.........months
5.5 What is the first supplementary food given to the child?	

5.6 What kind of foods (including beverages) are given to the child in addition to breast milk and by what method of feeding? [x] code cs: cup and spoon c: cup b: bottle (feeding bottle) h: hand ch: chewing by mother (Go to question 5.11 or 5.12)	food	method of feeding[x]
5.7 If not, at which age of the child did you discontinue breast-feeding?	.......months	
5.8 Did you stop breast-feeding • At once? • Gradually?	[ ] [ ]	
5.9 How did you stop breast-feeding?		
5.10 Why did you stop breast-feeding?		
5.11 What kind of foods do you give to your child?		
5.12 Which of the foods given to the child are prepared at home or bought specially for the child?	Home prepared infant foods Purchased infant foods	
5.13 Which foods given to the child are just ordinary adult foods?	Adult foods given to infants	

B6 Food avoidances or taboos

(Information from head of the household and if male also from his wife)

Generally, food avoidances observed by the head of the household and his wife are decisive for the food habits of the household. If you are interested in food avoidances of the old and young generations, other members of the household can be interviewed as well. Fill in the name and code number of the respondents and record not only foods that should not be eaten but also the reason why.

Place:.......... Address:........ Name fieldworker:.................. Date:...............

Signature of supervisor: ....

Identification no. of household: ....

Name of head of household: ....

Name and person number of the respondents				
Which foods and drinks may not be consumed by the following categories and why?	Kinds of food and drink	Reason	Kinds of food and drink	Reason
• Infants during weaning				
• Girls				
• Women during monthly periods				
• Women in general				
• Men in general				
• Sick persons				

B7 Special foods

(Information from head of the household and if male also from his wife)

Question 7.1 deals with special foods for pregnant and lactating women; Questions 7.2 and 7.3 with food and drinks for special occasions; Question 7.4 with special kinds of food or unusual substances such as bark or edible earth, consumed during food shortages and famine.

Place:.......... Address:........ Name fieldworker:................. Date:..............

Signature of supervisor: ....

Identification no. of household: ....

Name of head of household: ....

Name and person number of respondents		
7.1 Do women have special foods and drinks during • pregnancy? • lactation?	Kinds of food and drink	Kinds of food and drink
7.2 What kind of food and drink are used for		
• Celebrating birth of a child		
• Puberty rites, boys?		
• Puberty rites, girls?		
• Weddings?		
• Funerals		
• Sowing or planting ceremonies?		
• Harvest ceremonies?		
• Other (stated)		
....		
7.3 What kind of food and drink does one offer to distinguished guests?		
7.4 What kind of foods are consumed during food shortages and famine?		

B8 Frequency of food consumption

Collect the data on food consumption from a selected number of the target population. The form is based on the 24-hour recall method. The list of foods printed on the form should be extended if information is already available. For larger households, two sheets will be necessary in order to interview more than four members. Ask each member of the household, "What kinds of food and drink did you eat yesterday?". For children, the mother or responsible person should be asked.

Unless otherwise required, collect the data on a "normal" day and not on a feast day or other special occasion. In surveys on food habits respondents are usually asked to report only on the kind (qualitative survey) and not the amount of food consumed. The collected data represent the dietary pattern of the household and its members. Accurate quantitative information is difficult and time consuming to collect by means of the recall method.

If there is a pattern of three meals daily, the data should be obtained separately for the morning meal (M), noon meal (N), and evening meal (E). A three-meal pattern is not universal. Place a cross in the heading M, N, or E when that meal is the main one of the day. Snacks can be recorded by including them in the nearest meal, or by adding a new column to the questionnaire concerning snacks. List all the foods consumed in the left column and tick them off under the column M, N, and E. Also state the origin of the food as bartered (B), home-produced (H), gift (G), or purchased (P). Describe the food consumed as completely as possible. For instance, do not write down only "cassava" or "meat", but also if it is dry or fresh, the types of meat or fish, whether smoked, roasted, or otherwise prepared. Dishes, such as stew should be listed, together with the ingredients.

Form B8 Frequencyof food consumption

Place:.......... Address:........ Name fieldworker:................. Date:..............
Signature of supervisor: ....

Identification no. of household: ....
Name of head of household: ....

What kind of foods did you eat yesterday? (24-hour recall) Day: . . .. Month:. . . . Year:. . . .

All members of the household (children included)

M: morning (breakfast)
N: noon (lunch)
E: evening (supper)

Origin code: (Or)
B: Barter
G: Gift
H: home-produced
O: Purchased

Name/person no. of respondent																
Kind of food	**M**	**N**	**E**	**Or**	**M**	**N**	**E**	**Or**	**M**	**N**	**E**	**Or**	**M**	**N**	**E**	**Or**
1 Cereals, roots, tubers (staple food)																
2 Grain legumes																
3 Vegetables/fruit																
4 Nuts																
5 Animal foods																
6 Fats/oils																

Form B8 Continued.

Name/person no. of respondent																
Kind of food	**M**	**N**	**E**	**Or**	**M**	**N**	**E**	**Or**	**M**	**N**	**E**	**Or**	**M**	**N**	**E**	**Or**
7 Beverages																
8 Salt																
9 Other foods																

Appendix C: Presentation of data on food ethnography: Some examples

In this section, some examples are given on presenting data in the form of tables not requiring complicated techniques. The tables presented here are obtained from a number of small-scale field studies in which the authors were involved.

Examples of how to present demographic, socio-economic data, and dietary patterns of a community are given in Tables C1-C21. There are of course several ways of presenting these data and the samples are only suggestions. Most of the tables give absolute figures. With large numbers it is useful to calculate the figures on a percentage basis. For more complex calculations and presentations it is necessary to seek for advice from a statistician. As was said in the introduction, all examples of this manual should be adapted to the specific aims and needs of the user. The tables should be presented and used in such a way that they support the written text of the study: findings and the discussion. Each table has to be provided with a clear title so that the table can be understood independently from the written text of the report or article.

Tables C1-C7 are suggestions for presenting the background of the research population, some demographic and socio-economic data. Table C3, for instance, indicates that only a few households in the community are female headed.

Examples of how to present an agricultural calendar and a list of consumed foods are presented in Tables C8 and C9. These tables are obtained from a food ethnography of the Otammari in north-western Benin, prepared by Van Liere *et al.* (1996). Table C10 is an example of a descriptive table, presenting a typical daily meal pattern.

The frequency of food consumption will give the dietary pattern of a community or group and is presented in Tables C11-C15. The dietary pattern is qualitatively expressed in terms of percentage and frequency of use of different foods. It gives the kind of food and not the quantity of food consumed. Thus, for instance in Table C11, an entry of 29 against "cassava" means that some "cassava" was eaten in 29 out of 100 meals. The average number of meals per day is calculated by dividing the total number of meals eaten by the number of persons. Very detailed information is given in Table C11. The same information can also be used to show the general dietary pattern, and to give information on dietary patterns of different groups such as adults - men and women - children and infants (Table C12).

Tables C15-C20 deal with dietary habits and infant feeding practices, C21 with consumption and attributes of key food items.

Table C1. Composition of the population of the community by age and sex.

Age (years)	Number	
	Male	Female
< 1	6	6
1 - 3	19	18
4 - 6	18	17
7 - 9	16	15
10 - 12	14	13
13 - 15	11	12
16 - 19	12	13
20 - 39	41	42
40 - 49	13	13
50 - 59	9	8
60 -69	5	6
> 70	3	5
< 1-70 +	167	168

Table C2. Absolute frequency of household size in the community.

Number of persons in household	Number of households
1 – 2	5
3 – 4	10
5 – 6	16
7 – 8	10
9 – 10	5
11 – 12	5
> 12	3
1 – 12 +	54

Table C3. Age and sex of heads of households in the community.

	Age (years)					
	20 - 29	30 - 39	40 - 49	50 -59	60+	20 – 60+
Male	6	8	9	6	19	48
Female	-	1	2	1	1	5
Total	6	9	11	7	20	53

Table C4. Educational level of boys and men in the community.

Age (years)	Illiterate or not attending school	Primary school only	Secondary school	Others	Total
4 - 6	9	9	-	-	18
7 - 9	2	14	-	-	16
10 - 12	1	9	4	-	14
13 - 15	1	4	6	-	11
16 - 19	4	3	5	-	12
20 - 39	10	28	3	-	41
40 - 49	10	-	2	1	13
50 - 59	7	1	1	-	9
60 >	7	-	1	-	9
4 – 60+	51	68	22	1	142

Table C5. Educational level of girls and women in the community.

Age (years)	Illiterate or not attending school	Primary school only	Secondary school	Others	Total
4 - 6	9	8	-	-	17
7 - 9	6	9	-	-	15
10 - 12	1	9	-	-	13
13 - 15	3	3	6	-	12
16 - 19	4	3	6	-	13
20 - 39	13	26	2	1	42
40 - 49	12	1	-	-	13
50 - 59	7	1	-	-	8
60 >	10	1	-	-	11
4 – 60+	65	61	17	1	144

Table C6. Main and subsidiary occupation of men in the community, 16 years and older.

Main occupation	Number	Subsidiary occupation				
		farmer	trader	artisan	other	none
Farmer	50	+	4	5	2	39
Trader	8	5	+	-	2	1
Artisan	4	3	-	+	-	1
Other	2	-	-	-	+	2
Unemployed	19	1	-	-	-	18
Total	83	9	4	5	4	61

Table C7. Occupation of women in the community, 16 years and older.

Occupation	Number
Farmer	60
Trader	2
Artisan	3
Other	1
Home-maker only	15
None	6
Total	87

Table C8. Agricultural calendar of the Otammari in the commune of Manta, north-western Benin.

	Dry season				Rainy season					Dry season		
Crops **Months**	**J**	**F**	**M**	**A**	**M**	**J**	**J**	**A**	**S**	**O**	**N**	**D**
EARLY BEANS *Phaeseolus lunatus*					S		H	H				
YAM *Dioscorea spp.*		S	S					H	H	H	H	H
MAIZE *Zea mays*					S	S		H				
EARLY MILLET *Pennisetum spp.*						S		H				
FONIO *Digitaria exilis*					S	S			H	H		
RICE *Oryza sativa*					S	S				H	H	
BEANS *Vigna spp.*						S				H		
SWEET POTATO *Ipomoea batatas*						S				H	H	
BAMBARA GROUNDNUTS *Voandzeia subterranean*						S				H	H	
GROUNDNUTS *Arachis hypogaea*					S	S				H	H	
SORGHUM *Sorgum spp.*					S	S					H	H
MILLET *Pennisetum spp.*					S	S					H	H

S = Sowing or planting
H = Harvest
Source: Van Liere, 1993

Table C9. List of vegetable products consumed by the Otammari in north-western Benin.

English name	French name	Scientific name	Ditammari name
Cereals	Céréales		
Sorghum	Sorgho	*Sorghum spp.*	eyonou
Millet	Petit Mil	*Pennisetum spp.*	eyomata, inati
Hungry rice	Fonio	*Digitaria exilis*	ipoiga
Maize	Maïs	*Zea mays*	yemorio
Rice	Riz	*Oryza sativa*	imoua
Tubers	Tubercules		
Yam	Igname	*Dioscorea spp.*	yano
Sweet potato	Patate douce	*Ipomoea batatas*	yekyanyenra
Cassava	Manioc	*Manihot esculenta*	ifoga
Taro	Taro	*Colocasia esculenta*	yekonwo
Grain legumes	Légumineuses		
Beans	Haricots	*Vigna spp.*	yetoupia, isatou
Bambara groundnuts	Voandzou	*Voandzeia subterranea*	yogma
Groundnuts	Arachide	*Arachis hypogaea*	yekampian
Oil seeds and nuts	Noix oléagineux		
Cashew nut	Acajou	*Anacardium occidentale*	acajou
Pumpkin seeds	Courge	*Cucurbita spp.*	ia
Sesame seeds	Sésame	*Sesamum radiatum*	muwadomu
Baobab seeds	Baobab	*Adansonia digitata*	batotjebie
Groundnut	Arachide	*Arachis hypogaea*	yekampian
Tiger nuts	Souchet	*Cyperus Esculentus*	esantetikwan
Vegetables	Légumes		
Tomatoes	Tomate	*Lycopersicon esculentum*	yaperko
Okra	Gombo	*Hibiscus esculentus*	yanoura
Onions	Oignons	*Allium spp.*	
Squash	Courge	*Cucurbita esculentum*	yeka
Green leaves			
amaranth	Amaranthe	*Amaranthus cadalis*	
okra	Gombo	*Hibiscus esculentus*	tinoufant
baobab	Baobab	*Adansonia digitata*	titenakanti
squash	Courge	*Cucurbita esculentum*	beyambie
bitter leaves	Feuilles amères	*Vernonia amygdalina*	tikounteti
African eggplant	Aubergine	*Solanum macrocarpon*	tikanfati
Wild custard apple	Pomme canella	*Annona senegalensis*	imouti
Red flower cotton tree	Kapokier	*Bombax buonopozense*	yafoga
Monkey guave	Ebène de marais	*Diospyros mespiliformis*	yapi

Table C9. Continued.

English name	French name	Scientific name	Ditammari name
Fruits	Fruits		
Mango	Mangue	*Mangifera indica*	yapeta
Citrus fruit	Citrus	*Citrus spp.*	dieml
Shea fruit	Karité	*Butyrospermum parkii*	ditani
Borassus palm fruit	Fruit de rônier	*Borassus aethiopicum*	dityeteri
African oak fruit	Prunier noir	*Vitex cienkowskii*	dimantoni
Fats	Graisse		
Shea butter	Karité	*Butyrospermum parkii*	fakouafa
Groundnut oil	Arachide	*Arachis hypogaea*	yekampian
Condiments	Condiments		
African locust (mustard)	Néré	*Parkia biglobosa*	ii-jou
Red peppers	Piment	*Capsicum annuum*	ikoudani

Source: Van Liere, 1993.

Table C10. Typical meal pattern of the Otammari in north-western Benin.

Morning:	Warmed-up *pâte* (porridge), leftover from the evening before, of sorghum, or a mixture of hungry rice and finger millet. Relish of green leaves, mustard, hot pepper and salt.
Noon:	Freshly prepared *pâte* from sorghum; with a relish of okra, hot pepper, salt, mustard made from African locust seed, and shea butter.
Evening:	*Pâte* with a relish warmed-up from noon or a fresh prepared *pâte*.
Between meals:	Bambara groundnuts are roasted and consumed by children.
Beverages:	Water and local sorghum beer.

Source: based on Van Liere, 1993.

Table C11. Dietary pattern based on 100 meals in the community, 7 days, average 80 people observed at each meal.

	Relative frequency (%) of foodstuff			
	any meal	**breakfast**	**lunch**	**supper**
Proportion of meals at which item was recorded (%)	**87**	**90**	**80**	**90**
Plantain	2	-	0.7	6
Cassava	29	3	8	76
Yam	2	1	1	3
Cocoyam	6	0.1	1	17
Maize	48	63	64	13
Bread [wheat]	16	34	5	9
Doughnuts, biscuits	2	4	0.9	0.3
Rice	7	22	4	4
Tomatoes	76	56	80	94
Onions	75	55	79	94
Peppers	76	58	81	94
Aubergine	19	10	16	33
Okra	8	5	12	9
Leaves	1	0.3	4	0.3
Fruit	0.3	0.1	-	0.3
Beef	12	5	10	21
Bush meat	-	-	-	-
Pork	10	5	10	15
Mutton and goat meat	11	2	8	24
Smoked meat	3	0.8	2	7
Chicken	0.7	0.3	0.3	1
Snails	0.4	-	-	1
Hen eggs	0.2	0.3	0.3	0.2
Fresh fried fish	8	9	12	5
Smoked fish	43	23	40	68
Salted fish	4	3	9	2
Tinned fish	0.2	0.1	0.7	-
Shrimps, crabs	0.2	0.1	0.1	0.3
Milk [canned, condensed]	13	27	3	9
Groundnuts	65	2	4	11
Beans	8	21	1	0
Palm nuts	11	4	7	21
Palm oil	4	0.8	10	0.7

Table C11. Continued.

	Relative frequency (%) of foodstuff			
Proportion of meals at which item was recorded (%)	**any meal 87**	**breakfast 90**	**lunch 80**	**supper 90**
Palm kernel oil	1	1	2	-
Groundnut oil	0	-	-	0.1
Coconut oil	10	19	8	3
Sugar	22	46	9	12
Sweets	0.6	0.5	-	-
Beverages including tea and coffee	12	25	2	8
Spirits	3	3	1	3
Salt	80	69	82	93

Table C12. Dietary pattern on 100 meals of men (n = 42) and women (n = 38) in the community, August.

	Relative frequency (%) in meals	
	men	**women**
Roots or tubers	43	37
Maize	49	59
Other cereals	22	27
Sugar	19	24
Vegetables or fruits	82	77
Beans including groundnuts	13	13
Meat	31	28
Fish	60	60
Other animal products	12	14
Palm nuts or oil	11	11
Other oils and fats	13	11

Table C13. Dietary pattern of the community, classed as home-produced and purchased (including gifts), 80 people interviewed.

	Relative frequency (%) in meals	Proportion of foodstuffs (%)	
		home-produced	purchased
Roots and tubers	40	74	26
Maize	45	3	97
Bread	16	-	100
Rice	7	-	100
Beans including groundnuts	13	1	99
Palm nuts or oil	11	-	100
Tomatoes	76	82	18
Onions	75	1	99
Peppers	76	93	7
Meat	30	4	96
Fish	60	-	100

Table C14. Dietary pattern based on 100 meals by socio-economic classes in the community(n=80). The classes were defined as follows: high, big farmers and traders; middle, small farmers and artisans; low, landless labourers and squatters.

	Proportion of meals with food stuff (%)		
	High class	Middle class	Lower class
Roots and tubers	32	40	48
Maize [wheat]	38	44	52
Bread	30	15	4
Rice	25	6	1
Pulses including groundnuts	20	13	10
Vegetable oil	25	10	8
Tomatoes	76	74	70
Onions	78	75	72
Peppers	78	76	71
Meat	40	28	5
Fish	60	55	20

Table C15. Seasonal influence on dietary pattern in the community (n = 80).

	Proportion of meals with the item (%)		
	March	June	October
Roots and tubers	2	4	32
Maize	76	22	48
Millet	0	72	17
Rice	18	8	0
Pulses including groundnuts	18	8	0
Leafy vegetables	16	9	0
Other vegetables and fruits	2	1	0
Meat	51	13	0
Fish	21	22	81

Table C16. Members of urban and rural households responsible for supplying money used for food purchase at the market.

Place	Husband	Wife	Husband and wife	Children	Relatives
Urban	9	7	3	1	-
Rural	11	3	4	1	-

Table C17. Frequency of cooking in urban and rural households.

Place	Once per day	Twice per day	Thrice per day
Urban	6	8	6
Rural	9	6	4

Table C18. Persons doing cooking in urban and rural households.

Place	Cooking alone	Cooking not alone					
		other wife	daughter	female relative	husband	other	total
Urban	6	-	6	5	1	2	14
Rural	6	2	8	2	1	-	11

Table C19. Duration of breast-feeding among urban and rural mothers.

Place	Duration in months					
	< 3	3 – 5	6 - 8	9 - 11	12 - 17	> 18
Urban	-	-	2	1	7	3
Rural	-	1	1	-	4	6

Table C20. Number of mothers giving supplementary foods to infants according to age and method of feeding in a rural community (n = 45).

	Age (months)						Method of feeding			
	3	3-5	6-8	9-11	12-24	>24	cup	spoon	feeding bottle	hand
Maize pap	1	16	20	-	3	-	4	29	1	6
Bread	-	-	2	-	-	-	-	2	-	-
Rice	-	-	2	-	1	-	-	-	-	-
Plantain (banana)	-	-	1	-	1	1	-	2	-	1
Yam	-	2	8	2	6	3	-	-	-	21
Cassava	-	-	1	-	2	1	-	-	-	4
Soup/stew	-	-	5	-	3	1	-	-	-	9
Fish	-	-	-	-	-	-	-	-	-	-
Meat	-	-	-	-	-	-	-	-	-	-
Milk (canned powdered)	-	6	3	-	-	-	2	3	4	-
Groundnuts	-	-	-	1	1	-	-	-	-	2
Hen eggs	-	-	1	-	1	-	-	-	-	2
Beans	-	-	3	-	-	-	-	1	-	2
Fruits	-	2	1	-	-	-	-	3	-	-
Beverages	-	-	3	-	-	-	-	3	-	-
Palm oil	-	1	4	1	2	-	X	X	X	X
Margarine	-	-	1	-	-	-	X	X	X	X

Table C21. Weekly consumption and attributes of key (pro)vitamin A rich food items consumed in Burkina Faso.

Food items	Children consuming (%)		Weekly frequency of consumption		Portion size (g)	RAE/ portion size	Key food attributes
	Dry season	Rainy season	Dry season	Rainy season			
Animal products							
Liver, ox	12.5	21	1.25±0.5	1.9±0.4	10.8±0.99[a]	1800	Strength and health provider, blood and vitamin rich food. Availability year-round and high cost, liver of especially hare is used to treat night blindness. Fried in oil, can be eaten by children.
Egg	3	0	1.0±0.0	na	25±0[b]	43.8	Strength health provider, blood rich food, vitamin rich food, Available 4 to 5 months a year, high cost, mostly used for breeding or sale to get cash. Fried in oil, can be eaten by children.
Cow milk	0	21.2	na	2.3±0.75	100±1.41[c]	27	Strength health provider, blood and vitamin rich food. Availability 3 to 4 months and high cost. Can be eaten by children.
Fruits and tuber							
Mango	71.9	0	5.1±6.5	na	147.5±83.2	224.3	Strength and health provider, blood and Vitamin rich food. Available 5 to 6 months a year and low cost. Snack, fills your stomach.

Papaya	6	0	1.0±0.0	na	125	31.3	Strength health provider, blood and vitamin rich food. Available one to two months, high cost, mostly sale for cash. Can be eaten by children.
Yellow flesh sweet potato	0	0	na	na	na	na	Strength health provider, blood and vitamin rich food. Available one month in small quantity. Should not be consumed everyday by children. Fried in oil, boiled or eaten raw snacks
Vegetables							
Boundo leaves *Ceratotheca sesamoids*	0	66.7	na	3.8±1.68	102.5±52.4[d]	13.3	Strength health provider, vitamin rich food. Fresh leaves available two to three months, sun dried and kept year round. Easy to digest.
Sorrel leaves *Hibiscus sabdarifa*	0	78.9	na	5. 5±2.69	134.5±74.4	88.8	Strength health provider, vitamin rich food. Available 6 to 7 months, affordable on wet season and grown for cash in dry season.
Boulvaka leaves *Corchorus olitorius/ tridens*	0	48.5	na	3.6±1.96	102.5±52.4[4]	30.6	Strength health provider, vitamin rich food. Available three to four months. Dried leaves not consumed.
Ben-oil-tree leaves *Moringa oleifera*	0	27.3	na	3.2±1.92	102.5±52.4	16.3	Strength health provider, vitamin rich food. Fresh leaves available three to four months. Not consumed, no recipe on how to process and cook it.
Amanrathus leaves *Amanrathus candilis*	0	24.2	na	2.0±1.31	102.5±52.4	17.9	Strength health provider, vitamin rich food. Fresh leaves available for three to four months, mostly grown for sale. Not consumed, no recipe on how to process.

Table C21. Continued.

Food items	Children consuming (%)		Weekly frequency of consumption		Portion size (g)	RAE/ portion size	Key food attributes
	Dry season	Rainy season	Dry season	Rainy season			
Bean leaves	0	42.4	na	3.6±1.55	102.5±52.4	9.3	Strength health provider, vitamin rich food Fresh leaves available three to four months. Fill your stomach, could be grown in family garden.
Baobab leaves *Adansonia digitata*	0	69.7	na	4.4±2.7	102.5±52.4	2.8	Strength health provider, vitamin rich food. Fresh leaves available two to three months, Sun dried and kept dried.
Yellow sorrel flower	0	0	na	na	na	na	Strength health provider, vitamin rich food. Fresh flowers available for tow months, sun dried and kept year round. Particular taste (acid).
Okra *Hibiscus esculentus*	100	0	8.9±7.2	na	102.5±52.4	3.6	Strength health provider, vitamin rich food, blood rich food. Fresh okra is available for 3 months, always processed and kept dried for year round consumption. Easy to cook, low cost.

[a] Average weight for liver bought at market.
[b] Only one child consumed egg.
[c] Portion in ml.
[d] No child consumed these vegetables during the 24-hours recall, so portion sizes could not be determined. Assumed is that portion sizes equal that of okra sauce.
Source: Nana *et al.*, 2005.

Appendix D:

Questionnaire on food consumption of an individual

For those working in food and nutrition programmes and who feel a need for data on food consumption, it is recommended to take the following aspects into account:

- sample size;
- time in days needed to record dietary intake;
- season of the year;
- measurement of the amount eaten by various individuals of the population sampled.

The example given here is a schedule for a 24-hour weighing record for an individual and has been used in a survey on food intake by school children.

The model consists of 3 sheets:
D1 General information;
D2 Record of food intake;
D3 Record of food preparation.

Sheet D3 and to some extent sheet D1 are necessary to calculate the nutrients supplied by cooked dishes with any precision.

D1 General information

List the names of household members with their age and sex. Indicate members eating particular meals outside the home as well as guests sharing any meal. The information on the questionnaire must be kept confidential, so remove names from the questionnaire as soon as possible. If more socio-economic information is needed, use the outline questions on food habits (Appendix B2 to B7), and the household members inventory sheet (Appendix B1.1 and B1.2).

Family name :
Address :
Respondent name :
Age :
Weight :
Interviewer :

General information

Code:
Date:

Persons in household by sex and age (specify children)

Men	Boys
Women	Girls

Morning meal	time of day	:
	absent	:
	visitors	:
Midday meal	time of day	:
	absent	:
	visitors	:
Evening meal	time of day	:
	absent	:
	visitors	:

D2 Record of a person's food intake in a day

- In recording foods, describe them carefully, for instance whether milk is whole or skimmed and whether flour is coarse or fine.
- Weigh dishes and containers before food is put on them.
- Where possible, weigh the food "as purchased" and the edible part (a check on your data).
- It may be impossible to weigh meals taken outside the home. Foods and dishes should then be reported in number or sizes in domestic measures such as cups, cigarette tins, bottles and by cost. Investigators must weigh samples to find out the average weight per measure or find out how much of the food can be bought from that same source for the money spent.

Family code :
Respondent name :
Date :

Time	Kind of food	Weight of empty plate (g) m_1	Weight of plate and food (g) m_2	Weight of plate and leftovers (g) m_3	Net intake (g) (m_2-m_3)	Remarks

D3 Food preparation by household

For cooked dishes, weigh:

- container or pan;
- raw ingredients;
- edible portion of each food;
- cooked food in container or pan;
- leftovers not eaten by any member of the household or guest during the day.

From this information and the amount of the cooked food eaten by the respondent, the amount of each ingredient eaten by the respondent can be calculated. The number of persons sharing the dish are a partial check on the actual amount eaten by the respondent.

Family code :
Respondent name :
Date :

Name of prepared food:

Weight of container pan:

Name of ingredient	Weight before preparation (g) m_1	Weight of inedible parts (g) m_2	Net weight raw (g) m_1-m_2	Remarks

Total
Weight of pan and cooked food
Weight of pan and leftovers
Weight of pan
Weight of food eaten

A meat sauce is taken as an example of how to calculate food intake.

(Example meat sauce) weight of pan: 500 g

Ingredients	Weight (g) 'as purchased'	Non edible part (g)	Net weight raw (g)
Tomatoes (sliced)	200	-	200
Onion (ground)	250	20	230
Minced fatty pork	50	-	50
Palm oil	35	-	35
Water	200	-	200
Red pepper	10	-	10

Total	1225
Weight of pan and cooked foods	1060
Weight of pan and leftovers	500
Weight of pan	500
Weight of food eaten	560

Assume the weight loss during cooking is due to the evaporation of water.

Persons sharing the dish: 1 adult male
2 adult females
2 children under 13 years of age

Respondent: boy 8 years of age
Amount eaten by the boy: 80 g
This seems a reasonable portion as judged by total weight of the dish and number of persons sharing the dish.

Assuming that the components are uniformly divided in the sauce, the amount of each ingredient eaten by the boy is as follows:

Fraction of sauce eaten by boy is 80 g / 560 g = 1/7
Intake of tomato = 200 g / 7 = 28 g
Intake of onion = 230 g / 7 = 33 g

Before calculating the nutrient value of prepared foods, check whether your food composition table gives data for cooked or raw foods.

D4 Instructions for collecting duplicate food samples

1. It is essential that you collect, as duplicates, exactly the same foods as you eat on……
   …(give day). Please buy enough food to make meals for an extra person (you will be reimbursed for extra costs).

2. It is easy to have a spare plate that is the same weight as the plate you are used to eat from.

3. Weigh out the food to eat and record the details (appendix D2).

4. On to the spare plate, weigh out exactly the same quantities of food in the same order as before.

5. After you have eaten your meal, weigh the left over in the usual way and remove exactly the same quantities from the duplicate meal on the spare plate.

6. Put the duplicate into the container.

7. You do not have to collect, but you are asked to weigh and record as usual, the following:

- All soft and alcoholic drinks.
- Tea and coffee (added sugar and milk should go into the container).
- Water and mineral water.
- Other watery drinks; but milk and soup should go into the container.

8. If you are unable to collect a duplicate portion of some food please record it below:

- …….
- …….

9. Please keep your food container in the refrigerator.

Appendix E:
Examples of calculation of nutrient intakes

E1 How to do the worksheet

Table E1 is a 24-hour recall sheet for a girl 6 years old in a commune in rural North Vietnam.

Table E2 shows the weight of the foods indicated in table E1. The 24-hour recall method uses household measures (cups, bowls, *etc.*), which have to be converted into grams. This is done by repeated weighing of the household measures, as well as weighing food items on local markets and in local shops. The Vietnamese food composition table uses the values per 100 gram *raw* edible portion, therefore cooked/fried amounts need to be converted into raw amounts. This is done by weighing food items during cooking in several households. Furthermore information from the National Institute of Nutrition in Hanoi is used.

Table E3 converts food intake into intakes of major nutrients, iron, and vitamin C. Conversion was based on the Nutritive Composition Table of Vietnamese Foods (National Institute of Nutrition, 2000).

Data might be calculated with a computer, a calculator, or without any special equipment. There should be agreement beforehand on the rounding-off of data. Work with more numbers behind the decimal point than the food composition table indicates is useless and meaningless.

If, for instance, 25 g of food has been consumed and the food contained 19 g of fat per 100 g, it is obvious that the person obtained
$\frac{19}{100} \times 25 = 4.75$ g of fat.

It has become practice to round to the even integer preceding the 5. Thus in this case 5.0 g of fat. In the food record, the data also has been rounded off. Common practice is that amounts above the 10 grams are rounded to the nearest gram (without decimals). However, this rounding should be considered carefully. Other rules may need to be made for nutrient rich or energy-rich foods.

E2 Evaluation of the results

It is beyond the scope of this guide to explain how to use dietary standards on nutrient requirements. A few remarks on the interpretation of dietary survey data will suffice.

First, one cannot assess the girl's diet on the basis of one day (table El). Suppose we have 50 records for one day of schoolgirls 6 years of age, then we can evaluate their food consumption as a group and compare data with nutrient requirements laid down by experts.

Table E4 compares nutrient intake data from women in two different area's, a low income and a middle income area, of the city of Nairobi (Kenya). The women are divided into a consumer and a non-consumer group of non-home prepared foods ("street foods"). It is shown that the energy intake is higher in women from the middle income area, as well as the adequacy of the listed nutrients.

Table E5 compares the percentage of children consuming non-home prepared foods, "street foods", and the proportion of energy and nutrients provided by these street foods. The children are divided into pre-school, school-going and non school-going children, all living in a slum area of Nairobi (Kenya). There are no significant differences in the percentage of children consuming street foods and in the proportion of daily nutrient intake provided by street foods between the different groups of children.

Table E2. The weight of the foods as indicated in the 24-hour recall.

Amount	Grams cooked	Conversion factor cooked-raw	Grams raw
½ bowl rice	84	0.5	42
1 bowl rice	140	0.5	70
1 soup spoon fish sauce		-	7
1 soup spoon cooking oil		-	6
1 big size tofu		-	281
1 small bowl mustard greens	85	1.2	89
½ coffee spoon lard		-	8
1 full coffee spoon lard		-	17
1 small pick water spinach	6	1.05	6
1 small mandarin		-	93, waste 26%, edible amount = 69

Table E1. 24-hour recall of a six year old girl in Hien Quan commune, Phu Tho Province, North Vietnam.

24-hour recall

Child ID: ß5023
Date: 2004-10-19
People living in the household: 2 adults, 3 children

Meal	Place	Dish	Ingredients	Preparation	Amount	Measure/size	Participants		Portion eaten by child
							Adults	Children	
Breakfast	home	Fried rice	Ordinary rice	cooked	½	Small bowl		1	1
			Fish sauce		1	Soup spoon		1	1
			Cooking oil		1	Soup spoon		1	1
Lunch	home	Rice with tofu	Ordinary rice	cooked	½	Small bowl		1	2
			Tofu	cooked	1 big size tofu	10 pieces	2	3	2 pieces
		Vegetable soup	Mustard greens	cooked	1	Small bowl	2	3	1/5
			Lard		½	Coffee spoon	2	3	1/5
Snack	home		Mandarin	peeled	1	Small size		1	1
Dinner	home	Rice	Ordinary rice	cooked	1	Small bowl		1	1
		Vegetables	Water spinach	cooked	3	Small 'picks'		1	1
		Meat	Lean pork meat	fried	100 grams	10 pieces	2	3	2 pieces
			Fish sauce		2	Soup spoons	2	3	
			Lard		1	Full coffeespoon	2	3	

Source: Van Wijngaarden, 2005.

Table E3. Calculation of nutrient intakes.

Nutrient intake, as calculated from the 24-hour recall in Table E1.

Meal	Food	Raw amount (g)	Energy (kcal)	Fat (g)	Protein (g)	Carbohydrate (g)	Iron (mg)	Vit C (mg)
Breakfast	Ordinary rice	42	145	0.4	3.3	32.0	0.5	-
	Fish sauce	7	2	-	0.5	-	0.2	-
	Cooking oil	6	57	6.3	-	-	-	-
Lunch	Ordinary rice	84	290	0.8	6.6	64	1.0	-
	Tofu	56	53	3.0	6.1	0.4	1.2	-
	Mustard greens	18	3	-	0.3	0.4	0.3	4.6[a]
	Lard	2	15	1.7	-	-	0.0	-
Snack	Mandarin	69	26	-	0.6	5.9	0.3	38.0
Dinner	Ordinary rice	70	241	0.7	5.5	53.4	0.9	-
	Water spinach	19	4	-	0.6	0.5	0.3	2.2[a]
	Lean pork meat	20	28	1.4	3.8	-	0.2	-
	Fish sauce	3	1	-	0.2	-	0.1	-
	Lard	4	31	3.5	-	-	0.0	-
Total		400	896	17.8	27.5	156.6	5	

[a] vitamin C loss during cooking/ blanching estimated as 50% (Bergstroem, 1994; Yadav and Sehgal, 1995)

Percentage of energy derived from protein, fat and carbohydrate

			Energy %
Protein:	27.5 x 4 =	110.0 = (462 kJ)	12
Fat:	17.8 x 9 =	160.2 = (769 kJ)	18
Carbohydrate:	156.6 x 4 =	626.4 = (3007 kJ)	70
Total:		896 = (4300 kJ)	

Table E4. Energy intake and % from protein and fat (mean ± SD) and adequacy of energy and nutrients (percentage of RDA presented as: median [25th – 75th percentiles]) according to study area and consumption or non-consumption of non-home prepared foods for women in Nairobi, Kenya.

	Korogocho slum area		**Dandora** Low-middle-income area	
Characteristics	**Consumers**	**non-consumers**	**consumers**	**non-consumers**
n	94	38	97	25
Energy intake in MJ[a]	5.1± 0.9	5.2 ±1.7	6.7 ± 1.9	6.5 ± 1.9
Energy% from protein[a]	9.6 ± 1.7	10.1 ± 1.9	9.7 ± 1.5	10.2 ± 1.4
Energy% from fat [a]	18.3 ± 5.7	16.9 ± 6.2	24.6 ± 6.2[b]	21.4 ± 5.6
Adequacy (%)				
Energy[a]	61 [51-68]	61 [47-68]	77 [62-92]	71 [60-90]
Protein[a]	65 [57-85]	70 [56-87]	92 [72-111]	86 [74-109]
Vitamin A[a]	34 [22-46]	36 [24-49]	60 [43-79]	64 [48-87]
Iron [a]	124 [110-161]	118 [97-169]	207 [151-305]	174 [144-241]
Calcium[a]	136 [103-183]	134 [89-169]	167 [136-206]	170 [139-211]

[a]Difference between Korogocho and Dandora, P<0.001
[b]Difference between consumers and non-consumers in the area, P<0.05
Source: Van 't Riet, 2002.

Table E5. The percentage of children consuming street foods, energy percentage from protein and fat in street foods and the proportion of energy and nutrients provided by street foods in the diet of those 5-7 year old children, living in a slum area of Nairobi, Kenya.

	Pre-school	School going	Non school going	All
% consuming street foods	73	73	89	78
N	27	29	33	89
Energy% from protein in sf[a]	11.8 ± 0.8	11.1 ± 0.8	11.7 ± 0.9	11.5 ± 0.5
Energy% from fat in sf[a]	26.8 ± 2.5	32.5 ± 3.4	24.7 ± 1.8	27.9 ± 1.5
Proportion (%) of daily intake provided by street foods (median [25th – 75th percentile])				
Energy	24 [14-36]	27 [15-36]	17 [9-30]	21 [13-32]
Protein	26 [16-36][b]	29 [15-39]	20 [10-31][b]	24 [15-36][c]
Fat	31 [20-55][c]	39 [22-58][c]	30 [13-48][c]	35 [18-55][c]
Vitamin A	13 [3-35]	23 [6-47]	6 [2-25] [c]	13 [3-34][c]
Iron	18 [11-29]b	19 [11-30][b]	14 [5-26]	17 [9-28]c
Calcium	19 [11-34]	27 [9-39]	20 [9-31]	22 [10-35]

[a] mean ± SE
Different from portion of daily intake by energy
[b] P<0.05
[c] P<0.01
Source: Van 't Riet, 2002.

Table E6. Application of dietary reference intakes for healthy individuals and groups.

Type of use	For the individual	For a group
Planning	RDA: aim for this intake	EAR: use in conjunction with a measure of variability of the group's intake to set goals for the intake of a specific group.
	AI: use as a guide for intake	AI: use as tentative goal, when an EAR is unavailable.
		UL: use to ensure that goals for the mean specific population groups do not place individuals within the group at risk.
Assessment	EAR: to examine the possibility of inadequacy. UL; to examine possibility of over consumption.	EAR: to examine the prevalence of inadequacy within a group

EAR Estimated average requirement
RDA Recommended dietary allowances
AI Adequate intake
UL Tolerable upper intake level.
Source: based on IOM Dietary reference Intakes 2000 (Food-and-Nutrition-Board, 2000, 2005).

Appendix F:
Anthropometric data collection

F.1. Child height measurement

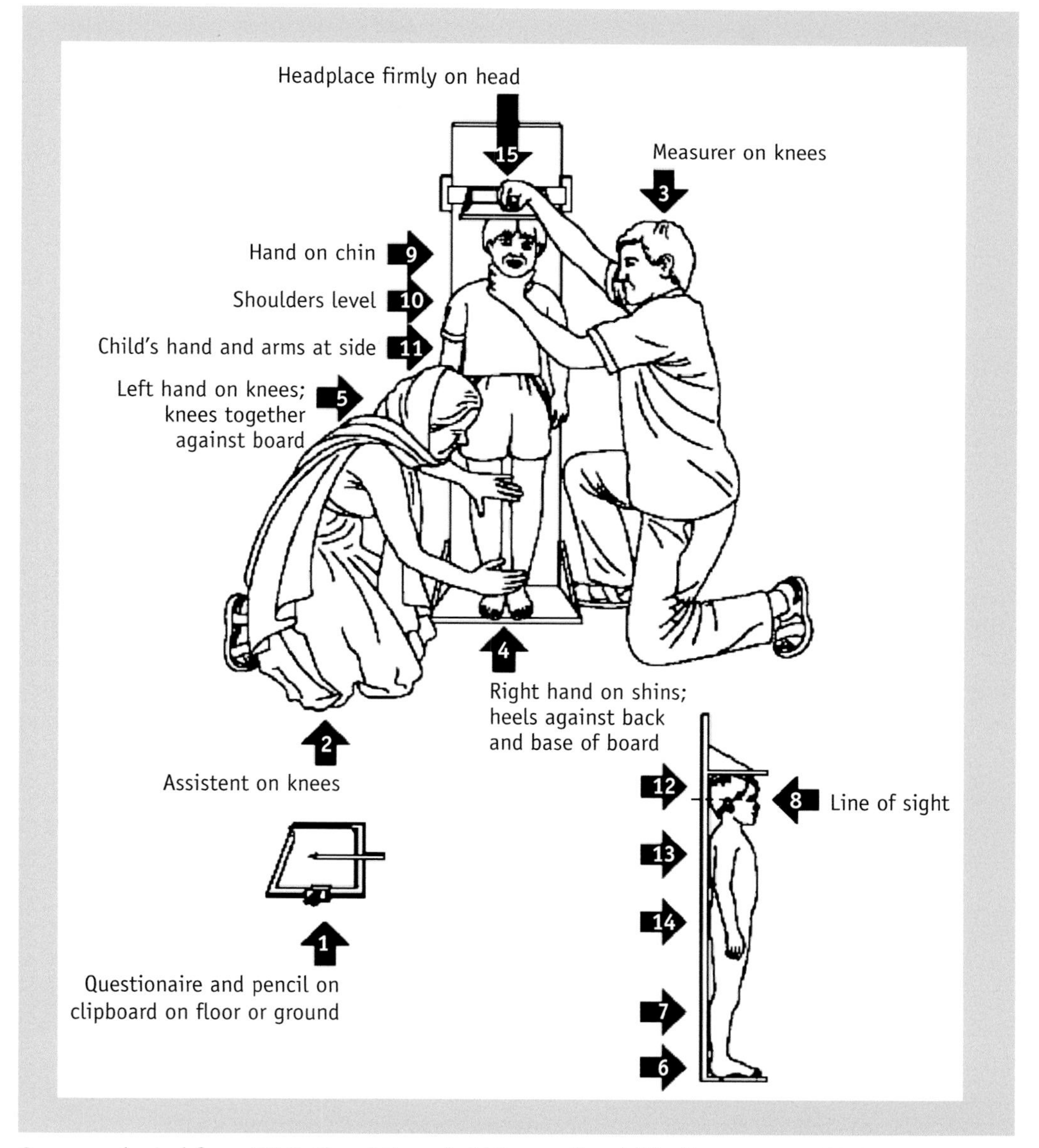

Source: adapted from UN National Household Survey Capability Programme, 1986.

F.2. Child length measurement

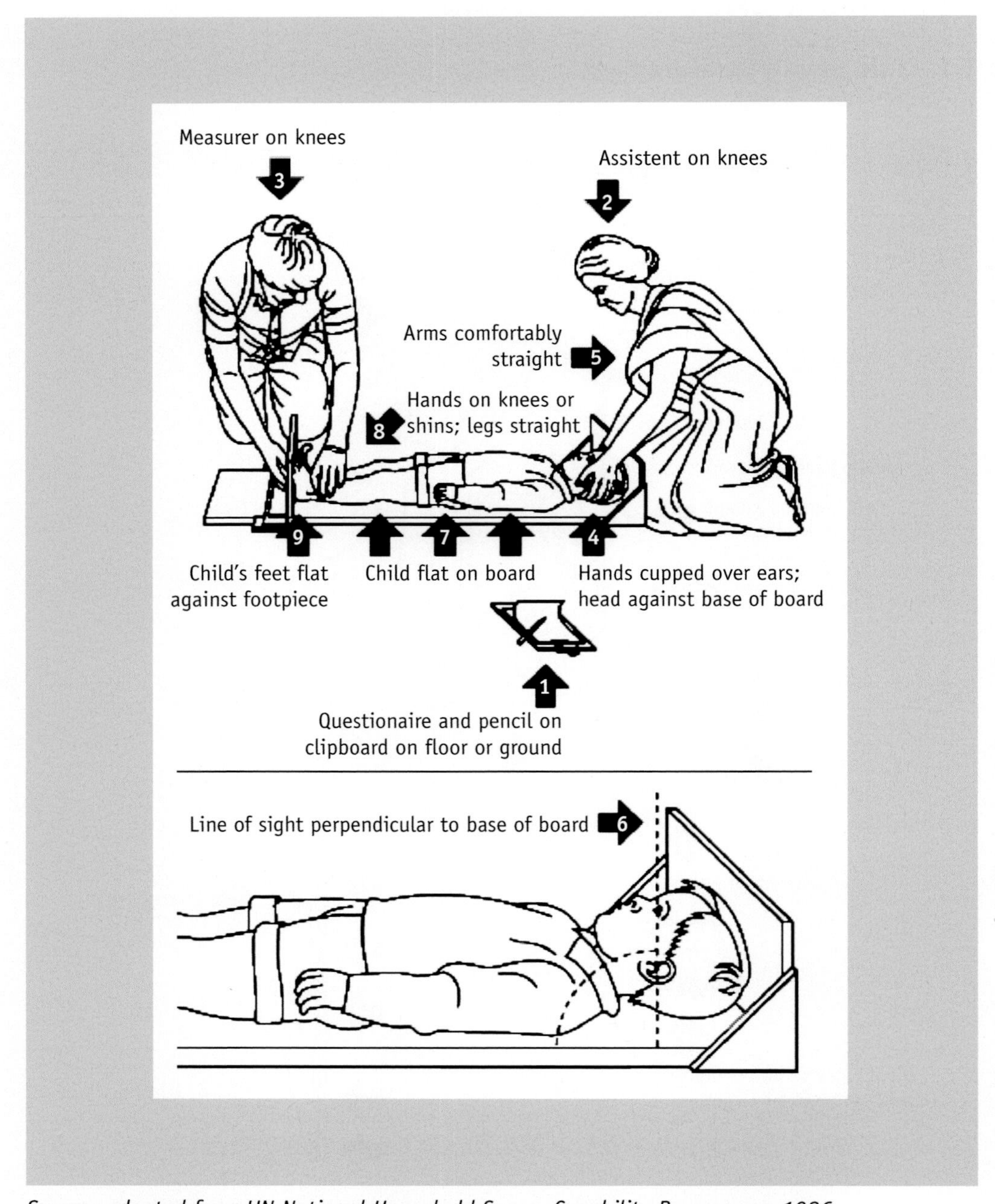

Source: adapted from UN National Household Survey Capability Programme, 1986.

F.3. Example data collection sheet anthropometry (from de Groot and Van Staveren, 1988)

Indentification number |_|_|_|_|_|_|_|_|_|_|_|_|

Interviewer code |_|_|_|

Date of measurement |_|_|_|_|_|_|
dd/mm/yy

1. Are the anthropometric measurements taken in the morning, after breakfast?
   [ ] 1. yes
   [ ] 2. no, at another time of the day
   [ ] 3. no, all anthropometric measurements are missing (continue with Q7) |_|
2. Height (10^{-1} cm) |_|_|_|•_|
3. Weight (10^{-1} kg) |_|_|_|•_|
4. Skinfolds (10^{-1} mm)
   a. triceps |_|_|•_|
   |_|_|•_|
   |_|_|•_| Average: |_|_|•_|
   b. biceps |_|_|•_|
   |_|_|•_|
   |_|_|•_| Average: |_|_|•_|
5. Circumference (10^{-1} cm)
   a. upper arm |_|_| •_|
   |_|_| •_| Average: |_|_| •_|
   b. waist |_|_| •_|
   |_|_| •_| Average: |_|_| •_|
   c. hip |_|_| •_|
   |_|_| •_| Average |_|_| •_|
6. Has the subject been measured in his/her under garment?
   [ ] 1. yes
   [ ] 2. no
   [ ] 8 irrelevant |_|
   If no, did you correct the circumferences?
   [ ] 1. yes
   [ ] 2. no
   [ ] 3. irrelevant |_|
7. If the anthropometric measures are missing, what was the reason?
   [ ] 1. subject refused
   [ ] 2. subject could not be measured due to illness, confinement to bed
   [ ] 3. subject did not show up/was not at home
   [ ] 4. other reasons
   [ ] 8. irrelevant |_|

Websites and references

Relevant websites on food and nutrition

www.aed.org	Academy for Educational Development, Washington
www.fao.org	Food and Agriculture Organization of the UN, Rome
www.ifpri.org	International Food Policy Research Institute, Washington
www.ivacg.org	International Vitamin A Consultative Group, Washington
www.unicef.org	UNICEF, New York
www.who.org	World Health Organization, Geneva

Recommended reading

Latham, M. C. (2004). *Human nutrition in the developing world.* Rome, FAO.

FAO (1999 - onwards). The state of food insecurity in the World. *The state of food insecurity in the World (SOFI).* Rome, FAO.

Geisller, C. and H. Powers, Eds. (2005). *Human nutrition.* 11th edition. London, Elsevier Limited.

Gibson, R. S. (2005). *Principles of nutritional assessment.* 2nd edition. New York, Oxford University Press.

Greenfield, H., D. A. T. Southgate, B. A. Burlingame and U. R. Charrondiere (2003). *Food composition data, production management and use.* 2nd edition. Rome, FAO.

Katz, S. H. and W. W. Weaver, Eds. (2003). *Encyclopedia of food and culture.* New York, Charles Scribner's Sons.

Kiple, K. F. and K. C. Ornelas, Eds. (2000). *The Cambridge world history of food.* New York, Cambridge University Press.

Vaughan, J. G., C. A. Geissler, B. E. Nicholson, E. Dowe and E. Rice (1997). *The new Oxford book of food plants* Oxford, Oxford University Press.

References

Anonymous (1985). *The origin and domestication of cultivated plants: symposium organized by Centro Linceo Interdisciplinaire di Scienze Matematiche e Loro Applicazioni, Academia Nazionale dei Lincei, Rome, 25 - 27 November, 1985,* Rome, Elsevier.

Anonymous (2002). World Food Summit five years later. *Urban Agriculture Magazine.*

Anonymous (2004). HIV/AIDS: a guide for nutrition, care and support. from http://www.fantaproject.org/downloads/pdfs/HIVAIDS_Guide02.pdf.

Anonymous (2006). Uniform Requirements for Manuscripts Submitted to Biomedical Journals: Writing and Editing for Biomedical Publication. Philadelphia, International Committee of Medical Journal Editors.

Abraham, A. L. (1986). Fire and cooking as a major influence on human cultural advancement: An anthropological/botanical/nutritional perspective. *Journal of Applied Nutrition* 38: 24-29.

Adandé, A. S. (1984). *Le mais et ses usages au Bénin méridional.* Dakar, Les Nouvelles Editions Africaines.

Ahmed Ali (2005). *Livelihood and food security in rural Bangladesh. The role of social capital.* Wageningen, Wageningen University. PhD thesis: 263.

Aldana, M. (2004). A systematic assessment of how nutrition is addressed in Poverty Reduction Strategy Papers. Human Nutrition. Wageningen, Wageningen University. Msc thesis.

Arimond, A. and M. T. Ruel (2004). Dietary diversity is associated with child nutritional status: Evidence from 11 demographic and health surveys. *Journal of Nutrition* 134(10): 2579-2585.

Atkinson, S. J. (1992). Urban-rural comparisons of nutrition status in the Third World. *Food and Nutrition Bulletin* 14(4): 337-340.

Babu, S. C. and P. Pinstrup Andersen (1994). Food security and nutrition monitoring. A conceptual framework, issues and challenges. *Food Policy* 19(3): 218-233.

Balatibat, E. M. (2004). The linkages between food and nutrition security in lowland and coastal villages in the Philippines. Wageningen, Wageningen University. PhD thesis: 236.

Beaton, G. H., J. Burema and C. Ritenbaugh (1997). Errors in the interpretation of dietary assessments. *American Journal of Clinical Nutrition* 65(4): 1000S-1007S.

Benkheïra, M. H. (2000). *Islam et interdits alimentaires. Juguler l'animalités* Paris, Presses Universitaires de France.

Benkheïra, M. H. (2002). Tabou du porc et identité en Islam. *Histoire et identités alimentaires en Europe.* M. Bruegel and B. Laurioux. Paris, Hachette literatures: 37-51.

Bergstroem, L. (1994). *Nutrient losses and gains in the preparation of foods.* Uppsala, National Food Administration.

Besley, T. and R. Burgess (2000). *Does media make Government more responsive? Theory and evidence form Indian famine relief policy.* London, London School of Economics and Political Science.

Blum, L., J. Pertti, G. Pelto and H. V. Kuhnlein (1997). *Community assessment of natural food sources of Vitamin A. Guidelines for an ethnographic protocol.* Ottawa, International Development Research Centre.

Bottéro, J. (2002). *La plus vieille cuisine du monde.* Paris, Editions Louis Audibert.

Bourdieu, P. (1989). *Distinction. A social critique of the judgment of taste.* London, Routledge.

Bricas, N. (1994). Street foods and trends in urban eating patterns. *Children in the Tropics* 213: 36-39.

Brillat-Savarin, J. (1825/1984). *The philosopher in the kitchen.* Hardmonthsworth, Penguin Book.

Brouwer, I. D. (1994). Food and fuel. A hidden dimension in human nutrition. Human Nutrition. Wageningen, Wageningen University. PhD thesis: 192.

Brouwer, I. D., A. P. den Hartog, M. O. K. Kamwendo and M. W. O. Heldens (1996). Wood quality and wood preferences in relation to food preparation and diet composition in Central Malawi. *Ecology of food and Nutrition* 35: 1-13.

Bukkens, S. G. F. (1997). The nutritional value of edible insects. *Ecology of Food and Nutrition* 36: 287-319.

Burgess, A. and R. F. A. Dean, Eds. (1962). *Malnutrition and food habits, report on an international and inter-professional conference.* London, Tavistock.

Burke, B. S. (1947). The dietary history as a tool in research. *Journal of the American Dietetic Association* 23: 1041-1046.

Cade, J., R. Thompson, V. Burly and D. Warm (2002). Development, validation and utilization of food frequency questionnaires- a review. *Public Health Nutrition* 5: 567-587.

Cameron, M. E. and W. A. van Staveren, Eds. (1988). *Manuel on methodology of food consumption studies.* Oxford Medical Publication. Oxford, Oxford University Press.

Caplan, P. (2002). Food in middle class Madras households from the 1970s to the 1990s. *Asian food. The global and the local.* K. Cwiertka and B. Walraven. Richmond Surrey, Curzon Press: 46-62.

Cecelski, E. (1987). Energy and rural women's work: crisis, response and policy alternatives. *International Labour Review* 127: 41-63.

Chauliac, M. and P. Gerbouin - Rerolle (1994). Street food: a comprehensive approach. *Children in the Tropics* 213: 5-20.

Conway, J. M., L. A. Ingwersen, B. T. Vinyard and A. J. Moshfegh (2003). Effectiveness of US Department of Agriculture 5-step multiple-pass method in assessing food intake in obese and non-obese women. *American Journal of Clinical Nutrition* 77(5): 1171-1178.

Cordero Fernando, G. (1976). Table Exotica. *The culinary culture of the Philippines.* G. Cordero Fernando. Manila, Bancom Audiovision: 141-147.

Cornevin, R. (1981). *La République Populaire du Bénin. Des origines Dahoméennes à nos jours.* Paris, Editions Maisonneuve & Rose.

De Foliart, G. R. (1999). Insects as food: why the western attitude is important. *Annual Review of Entomology* 44: 21-50.

De Garine, I. (1972). The Socio-cultural Aspects of Nutrition. *Ecology of Food and Nutrition* 1: 143-163.

De Garine, I. and V. de Garine, Eds. (2001). *Drinking. Anthropological approaches.* New York, Berghahn Books.

De Garine, I. and G. Koppert (1988). Coping with seasonal fluctuations in food supply among the savannah populations: the Massa and Mussey of Chad and Cameroon. *Coping with uncertainty in food supply.* I. de Garine and G. A. Harisson. Oxford, Clarendon Press: 120-259.

De Onis, M. and M. Blössner (2003). The WHO global Database on Child growth and malnutrition: methodology and applications. *International Journal of Epidemiology* 32(4): 518-526.

De Onis, M., M. Blössner, E. Borgh, R. Morris and E. A. Frongill (2004). Methodology for estimating regional and global trends of malnutrition. *International Journal of Epidemiology* 33(6): 1260-1270.

Delavigne, A. E. (2002). Pas de cochon, pas de Danois. Viande de porc et identité Danoise, perspective anthropologique. *Histoire et identités alimentaires en Europe.* M. Bruegel and B. Laurioux. Paris, Literatures Hachette: 53-62.

Den Hartog, A. P. (1973). Unequal distribution of food within the household; a somewhat neglected aspect of food behaviour. *FAO Nutrition Newsletter* 10(b): 8-17.

Den Hartog, A. P. (2002). Acceptance of milk products in Southeast Asia. The case of Indonesia as a traditional non-dairying region. *Asian Food. The Global and the Local.* K. Cwiertka and B. Walraven. Richmond Surrey, Curzon Press: 34-45.

Den Hartog, A. P. and I. D. Brouwer (1990). Adjustment of food habits in situations of seasonality. *Seasons, food supply and nutrition in Africa.* D. W. J. Foeken and A. P. den Hartog. Leiden, African Studies Centre. 43: 76-88.

Dennis, W. (1973). *Children of the crèche.* New York, Appleton Century Crofts.

Deurenberg-Yap, M. and P. Deurenberg (2003). Is a Re-evaluation of WHO body mass index cut-off values needed? The case of Asians in Singapore. *Nutrition Reviews* 61(supplement 1): 80-87.

DeWalt, K. M. (1993). Nutrition and the commercialization of agriculture: ten years later. *Soc Sci Med* 36(11): 1407-16.

Diamond, J. (1999). *Guns, germs and steel. The future of human societies.* New York, Norton & Company.

Doling, A. (1996). *Vietnam on a plate. A culinary journey.* Hong Kong, Roundhouse Publications.

Douglas, M. (1966). *Purity and Danger: An Analysis of Concepts of Pollution and Taboos.* London, Routledge, K Paul.

Durnin, J. V. G., S. Drummond and K. Satyanarayana (1990). A collaborative EEC Study on seasonality and marginal nutrition. The Glasgow-Hyderabad study. *European Journal of Clinical Nutrition* 44(supplement 1): 19-29.

Economist (1996). What, no Big Mac? The first McDonald's in India. *The Economist* (October).

Engle, P. L. and L. Lhotska (1999). The role of care in programmatic actions for nutrition: designing programmes involving care. *Food and Nutrition Bulletin* 20(1): 121-135.

Engle, P. L., P. Menon and L. Haddad (1997). *Care and nutrition: concepts and measurement.* Washington DC, IFPRI.

FAO (1989). Street foods. Report of an FAO Expert Consultation, Jogjakarta, Indonesia, 5-9 December 1988. *FAO Food Nutr Pap* 46: 1-96.

FAO (1990a). *Roots, tubers, plantains and bananas in human nutrition.* Rome, FAO.

FAO (1990b). Street foods. Report of a FAO Expert Consultation Jogjakarta, Indonesia. 5-9 December 1988. *FAO Food and Nutrition Paper, 46.*

FAO (1992). *Maize in human nutrition.* Rome, FAO.

FAO (1995). *Le lait et les produits laitiers dans la nutrition humaine.* Rome, FAO.

FAO (1999 - onwards). The state of food insecurity in the World. *The state of food insecurity in the World (SOFI).* FAO. Rome, FAO.

FAO, Ed. (2004). *Globalization of food systems in developing countries: impact on food security and nutrition.* Food and Nutrition papers. Rome, Food and Agricultural Organization (FAO).

FAO/WHO/UNU (2004). Human Energy Requirements. FAO Food and Nutrition Technical Report. Report of a Joint FAO/WHO/UNU Expert Consultation. Rome, 17-24 October 2001. *Food and Nutrition Technical Report Series 1.* Rome, FAO.

Fidanza, F., Ed. (1991). *Nutritional Status assessment.* London, Chapman and Hall.

Flores, M. and M. Nelson (1988). Methods for data collection at household or institutional level. *Manual on food consumption studies.* M. E. Cameron and W. A. van Staveren. Oxford, Oxford University Press: 33-52.

Foeken, D. W. J. and A. P. den Hartog, Eds. (1990). *Seasons, food supply and nutrition in Africa.* Research Reports. Leiden, African Studies Center.

Foeken, D.W.J., M. Sofer and M. Mlozi (2004). *Urban agriculture in Tanzania: issues of sustainability.* Leiden, African Studies Center, Research Report 75.

Fokkema, C. M. and G. Groenewold (2003). De migrant als suikeroom (the migrant as rich uncle). *Demos - Nidi* 19: 45-47.

Food-and-Nutrition-Board. (2000, 2005). Dietary Reference Intake. from www.iom.edu.

Frankenberger, T. R., R. M. Caldwell and J. Mazzeo (2002). Empowerment and governance: basic elements for improving nutritional outcomes. *Background paper for the 5th Report on the World Nutrition Situation.* Geneva, SCN.

Fumey, G. and O. Etcheverria (2004). *Atlas mondial des cuisines et gastronomie. Une géographie gourmande.* Paris, Editions Autrement.

Gade, D. W. (1976). Horsemeat as Human Food in France. *Ecology of Food and Nutrition* 5: 1-11.

Garrett, L. (2005). The next Pandemic? *Foreign Affairs* 84(4): 3-23.

Gibson, R. S. (2005). *Principles of nutritional assessment.* 2nd edition. New York, Oxford University Press.

Gillespie, S. and S. Kadiyala. (2005). HIV/AIDS and food and nutrition security. From evidence to action. *Food Policy Review.* from http://www.ifpri.org/pubs/fpreview/pv07/pv07.pdf.

Gomez, F., R. R. Galvan, J. Cravioto and S. Frenk (1955). Malnutrition in infant and childhood. *Advances in Pediatrics* S. Z. Levine. New York, Yearbook publishers. VII.

Goody, J. (1982). *Cooking, cuisine and class: a study in comparative sociology.* Cambridge, Cambridge University Press.

Greenfield, H., D. A. T. Southgate, B. A. Burlingame and U. R. Charrondiere (2003). *Food composition data, production management and use.* 2nd edition. Rome, FAO.

Guthe, C. E. and M. Mead (1945). *Manual for the study of food habits.* Washington, National Academy of Sciences.

Haddad, L. and C. Geisller (2005). Food and nutrition policies and interventions. *Human Nutrition.* 11th edition. C. Geisller and H. Powers. London, Elsevier, Churchill Livingstone.

Harris, D. R. (1969). Agricultural systems, ecosystems and the origin of agriculture. *The domestication of plants and animals.* P. J. Ucko and S. M. Dimbly. London, Duckworth.

Harris, M. (1985). *Good to Eat. Riddles of Food and Culture.* New York, Simon and Schuster.

Heymsfield, S. B., C. McManus and J. Smith (1982). Anthropometric measurement of muscle mass: revised equations for calculating born-free muscle area.. *american Journal of Clinical Nutrition* 36(4): 680-690.

Hill, P. (1978). Food farming and migration from Fante villages. *Africa* 48: 220-230.

Hodge, A. M., J. Montgomery, G. K. Dowse, B. Mavo, T. Watt, M. Alpers and P. Zimmet (1996). Diet in urban Papua New Guinea populations with high levels cardiovascular risk factors. *Ecology of Food and Nutrition* 35: 311-324.

Hoogland, J.P., A. Jellema and W.M.F. Jongen (1998). Quality assurance systems. *Innovation of food production systems. Product quality and consumer acceptance.* W.M.F. Jongen and M.T.G. Meulenberg. Wageningen, Wageningen Pers: 139-158,

Hoopes, J. (1979). *Oral history: an introduction for students.* Chapel Hill, University of North Carolina.

Horton, D., P. Ballantyne, W. Peterson, B. Uribe, D. Gasapin and K. Sheridan (1993). *Monitoring and evaluating agricultural research: a sourcebook..* The Hague, CAB International, International Service for National Agricultural Research.

IFPRI (1998). *Urban challenges to food and nutrition security: a review of food security, health, and care giving in the cities.* Washington, IFPRI.

IFPRI (1999). *Are urban poverty and undernutrition growing? Some newly assembled evidence.* Washington DC, IFPRI.

Infoods. (2003). The international network of Food database systems. from http//www.fao.org/infoods/directory_en.stm.

Jiggins, J. (1986). Women and seasonality: coping with crisis and calamity. *IDS Bulletin* 17(3): 9-18.

Kaaks, R., P. Ferrari, A. Ciampi, M. Plummer and E. Riboli (2002). Uses and limitations of statistical accounting for random error correlations, in the validation of dietary questionnaire assessments. *Public Health Nutr* 5(6A): 969-76.

Karesh, W. B. and R. A. Cook (2005). The human animal link. *Foreign Affairs* 84(4): 38-50.

Katz, S. H. and W. W. Weaver, Eds. (2003). *Encyclopedia of food and culture.* New York, Charles Scribner's Sons.

Keasberry, I. N. (2002). Elder care, old age security and social change in rural Yogyakarta, Indonesia. Wageningen, Wageningen University. PhD thesis: 407.

Keller, W., G. Donoso and E. M. DeMaeyer (1976). Anthropometry in nutritional surveillance A review. *Nutrition Abstracts and Reviews* 46: 591-609.

Kelly, A., W. Becker and E. Helsing, Eds. (1991). *Food and Health Data: Their Use in Nutrition Policy-Making.* European Series, WHO Regional Publications.

Kennedy, E. and T. Reardon (1994). Shift to non-traditional grains in the diet of East and West Africa: the role of women's opportunity cost of time. *Food Policy* 19: 45-56.

Kennedy, G., G. Nantel, I. D. Brouwer and F. J. Kok (2006). Does living in an urban environment confer advantages for childhood nutritional status? Analysis of disparities in nutritional status by wealth and residence in Angola, Central African Republic and Senegal. *Public Health Nutr* 9(2): 187-93.

Kilara, A. and K.K. Lya (1992). Food and dietary habits of the Hindu. *Food Technology* 46: 94-104.

Kiple, K. F. and K. C. Ornelas, Eds. (2000). *The Cambridge world history of food.* New York, Cambridge University Press.

Kuhnlein, H. V. (2003). Micronutrient nutrition and traditional food systems of indigenous peoples. *Food, Nutrition and Agriculture* 32: 33-37.

Latham, M. C. (2004). *Human nutrition in the developing world.* Rome, FAO.

Lignon-Darmaillac, S. (2001). Les vignobles méditerranéens. *Questions de géographie. La Méditerranée.* V. Morinaux. Paris, Editions du Temps: 161-172.

Little, M. A., S. J. Gray and B. C. Campbell (2001). Milk consumption in African pastoral peoples. *Drinking. Anthropological approaches.* I. de Garine and V. de Garine. New York, Berghahn Books: 66-86.

Loevinsohn, M. and S. Gillespie. (2003). HIV/AIDS, food security and rural livelihoods: understanding and responding. *RENEWAL Working Paper No 2.* from http://www.isnar.cgiar.org/renewal/pdf/RENEWALWP2.pdf.

Lopez, E. and J. Munchnik (1997). *Petites entreprises et grands enjeux: le développement agroalimentaire locale.* Paris, L'Harmattan.

Maarmar, H. (2003). Les bestiaires sacrificiels. *Sacrifices en Islam. Espaces et temps d'un rituel.* A. M. Brisebarre. Paris, CNRS Editions: 246-282.

Margetts, B. M. and M. Nelson (1997). *Design Concepts in nutritional epidemiology.* 2nd edition. Oxford, Oxford University Press.

Marr, J. W. (1971). Individual dietary surveys: purposes and methods. *World Rev Nutr Diet* 13: 105-64.

Marsh, D. R., D. G. Schroeder, K. A. Dearden, J. Sternin and M. Sternin (2004). The power of positive deviance. *British Medical Journal* 329: 1177-1179.

Mauro, F. (1991). *Histoire du café.* Paris, Editions Desjonquères.

Mauss, M. (1947/ 1967). *Manuel d'Ethnographie.* Paris, Payot.

Maxwell, D. G. (1996). Measuring food insecurity: the frequency and severity of coping strategies. *Food Policy* 21: 291-303.

Mead, M. (1960). *Cultural change in relation to nutrition.* Malnutrition and food habits, Guernavaca, Mexico, Transtock Publications, London.

Mintz, S. W. (1985). *Sweetness and power; the place of sugar in modern history.* Harmondsworth, Penguin Books.

Molony, C. (1975). Systematic valance coding of Mexican hot–cold food. *Ecology of Food and Nutrition* 4: 67-74.

Mwangi, A. M. (2002). Nutritional, hygienic and socio-economic dimensions of street foods in urban areas: the case of Nairobi. Wageningen, Wageningen University. PhD thesis: 142.

Mwangi, A. M., A. P. den Hartog, D. W. J. Foeken, H. van 't Riet, R. K. N. Mwadime and W. A. van Staveren (2001). The ecology of street foods in Nairobi. *Ecology of Food and Nutrition* 40: 497-523.

Nana, C. P., I. D. Brouwer, N. M. Zagré and A. Traoré (2005). Community assessment of availability, consumption, and cultural acceptability of food resources of (pro)vitamin A: towards the development of a dietary intervention among preschool children in rural Burkina Faso. *Food and Nutrition Bulletin* 26(4): 356-365.

National Institute of Nutrition (NIN) and Ministry of Health, Eds. (2000). *Nutritive composition table of Vietnamese foods.* Hanoi, Medical Publishing House.

Nelson, M., A. E. Black, J. A. Morris and T. J. Cole (1989). Between- and within-subject variation in nutrient intake from infancy to old age: estimating the number of days required to rank dietary intakes with desired precision. *Am J Clin Nutr* 50(1): 155-67.

Niehof, A. (1999). Household, family, and nutrition research: writing a proposal. Wageningen, Household and Consumer Studies, Wageningen University.

Niehof, A. (2001). *Food and harmony in the Javanese Slametan.* Wageningen, Wageningen University.

Niehof, A. and L. Price (2001). Rural livelihood systems: a conceptual framework. *UPWARD Working paper series 5.* Wageningen, Wageningen University.

Oguntona, C. R. B. and T. O. Tella (1999). Street foods and dietary intakes of Nigerian urban market women. *International Journal of Food Sciences and Nutrition* 50: 383-390.

Panoff, M. and M. Perrin (1973). *Dictionnaire de L'Ethnologie.* Paris, Payot.

Pekkarinen, M. (1970). Methodology in the collection of food consumption data. *World Review of Nutrition and Dietetics* 12: 145-171.

Pelto, G. H., P. J. Pelto and E. Messer, Eds. (1989). *Research methods in nutritional anthropology.* Tokyo, United Nations University Press.

Pemberton, R. W. (2002). Wild-gathered foods as counter currents to dietary globalisation in South Korea. *Asian Food. The Global and the Local.* K. Cwiertka and B. Walraven. Richmond, Curzon Press: 76-94.

Pennartz, P. and A. Niehof (1999). *The domestic domain: changes, choices and strategies of family households.* Aldershot, Ashgate Publishing.

Pitte, J. R. (1991). *Gastronomie française. Histoire et géographie d'une passion.* Paris, Fayard.

Piwoz, E. G. and E. A. Preble. (2000). HIV/AIDS and nutrition: a review of the literature and recommendations for nutritional care and support in sub-Saharan Africa. from http://www.aed.org/ghpnpubs/publications/3-hiv-aids-nutrition-eng.pdf.

Quinn, V. J. (1994). Nutrition and national development. An evaluation of nutrition planning in Malawi from 1936 to 1990. Wageningen, Wageningen University. PhD thesis: 401.

Richards, A. I. and E. M. Widdowson (1936). A dietary study in N.E. Rhodesia. *Africa* 9: 166-196.

Roden, C. (2000). *New book of Middle Eastern food.* New York, Alfred A Knopf.

Rodriquez, A. S. (2001). Food and Nutrition surveillance systems. Wageningen, IAC.

Rogers, B. L. and N. P. Schlossman, Eds. (1990). *Intra-household resource allocation.* Tokyo, United Nations University Press.

Rudie, I. (1995). The significance of eating: cooperation, support, and reputation Kelatan Malay households. *Male and female in developing countries Southeast Asia* W. J. Karim. Oxford, Berg Publishers: 228.

Ruel, M. T. (2001). Can food-based strategies help reduce vitamin A and iron deficiencies? A review of recent evidence. *Food Policy Review 5.* Washington, IFPRI.

Sakr, A. H. (1975). Fasting in Islam. *J Am Diet Assoc* 67(1): 17-21.

Sanaka Arachi, R. B. (1998). Drought and household coping strategies among peasant communities in the dry zones of Sri Lanka. *Understanding vulnerability: South Asian Perspective.* J. Twigg and M. R. Bhatt. London, Intermediate Technology Publications: 27-47.

Sayer, J. and B. Campbell (2003). *The science of sustainable development.* Cambridge, Cambridge University Press.

SCN, Ed. (2004). *Nutrition for improved development outcomes. 5th Report on the world nutrition situation.* Geneva, SCN.

Scrimshaw, N. S. and E. B. Murray (1988). The acceptability of milk and milk products in populations with a high prevalence of lactose intolerance. *Am J Clin Nutr* 48(4 Suppl): 1079-159.

Seidell, J. C. and T. L. Visscher (2000). Body weight and weight change and their health implications for the elderly. *Eur J Clin Nutr* 54 Suppl 3: S33-9.

Serra-Majem, L., D. MacLean, L. Ribas, D. Brule, W. Sekula, R. Prattala, R. Garcia-Closas, A. Yngve, M. Lalonde and A. Petrasovits (2003). Comparative analysis of nutrition data from national, household, and individual levels: results from a WHO-CINDI collaborative project in Canada, Finland, Poland, and Spain. *J Epidemiol Community Health* 57(1): 74-80.

Shack, W. A. (1978). Anthropology and the Diet of Man. *Diet of Man, Needs and Wants.* J. Yudkin. London, Applied Sciences Publishers: 261-280.

Shaxson, A., P. Dickson and J. Walker (1979). *The Malawi cookbook..* Zomba, Blantyre Printing and Publishing.

Shekar, M., J. P. Habicht and M. C. Latham (1991). Is positive deviance in growth simply the converse of negative deviance? *Food and Nutrition Bulletin* 13(1): 7-11.

Shetty, P. (2002). Food and nutrition: the global challenge. *Introduction to Human Nutrition.* M. J. Gibney, E. H. H. Vorster and F. J. Kok. Oxford, Blackwell publishing.

Simoons, F. J. (1973). The determinants of dairying and milk use in the old world: ecological, physiological, and cultural. *ecology of food and nutrition* 2: 83-90.

Simoons, F. J. (1994). *Eat Not This Flesh: Food Avoidances from Prehistory to Present.* Madison, University of Wisconsin Press.

Slimani, N., G. Deharveng, R. U. Charrondiere, A. L. van Kappel, M. C. Ocke, A. Welch, A. Lagiou, M. van Liere, A. Agudo, V. Pala, B. Brandstetter, C. Andren, C. Stripp, W. A. van Staveren and E. Riboli (1999). Structure of the standardized computerized 24-h diet recall interview used as reference method in the 22 centers participating in the EPIC project. European Prospective Investigation into Cancer and Nutrition. *Comput Methods Programs Biomed* 58(3): 251-66.

Smit, J., A. Ratta and J. NASr (1996). *Urban Agriculture: food, jobs and sustainable cities.* New York, UNDP.

Sujatha, T., V. Shatrugna, N. G. V. Rao, K. S. Padmavathi and P. Vidyasagar (1997). Street food: an important energy source for the urban worker. *Food and Nutrition Bulletin* 18(4): 318-322.

the Royal Anthropological Institute of Great Britain and Ireland, Ed. (1954). *Notes on queries on anthropology.* London, Rouledge and Kegan.

Thompson, F. E. and T. Byers (1994). Dietary assessment resource manual. *J Nutr* 124(11 Suppl): 2245S-2317S.

Tomkins, A. (2005). Evidence based nutrition interventions for the control of HIV/ AIDS. *South African Journal of Clinical Nutrition* 18(2): 187-191.

Trèche, S., A. P. den Hartog, M. J. Nout and A. Traoré (2002). Les petites industries alimentaires en Afrique de l'Ouest: situation actuelle et perspectives pour une alimentation saine. *Cahiers Agricultures* 11: 343-348.

Trichopoulou, A. and A. Naska (2003). European food availability databank based on household budget surveys: the Data Food Networking initiative. *Eur J Public Health* 13(3 Suppl): 24-8.

UN (1986). *National Household Survey Capability Programme. How to weigh and measure children. Assessing the nutritional status of young children in household surveys.* New York, United Nations.

UN (2000). World urbanization prospects: the 1999 revision. New York, UN Population Division.

UNICEF (1990). *Strategy for improved nutrition of children and women in developing countries.* New York, Unicef.

UNICEF (1994). *The urban poor and household food security. Policy and project lessons of how governments and the urban poor attempt to deal with household food security, poor health and malnutrition.* New York, Unicef.

Van 't Riet, H. (2002). The role of street foods in the diet of low income urban residents, the case of Nairobi. Wageningen, Wageningen University. PhD thesis: 117.

Van 't Riet, H., A. P. den Hartog, A. M. Mwangi, R. K. Mwadime, D. W. Foeken and W. A. van Staveren (2001). The role of street foods in the dietary pattern of two low-income groups in Nairobi. *Eur J Clin Nutr* 55(7): 562-70.

Van Huis, A. (2003). Insects as food in sub-Saharan Africa. *Insect Science and its Application* 32: 163-185.

Van Liere, M. J. (1993). Coping with household food insecurity: a longitudinal and seasonal study among the Otammari in North - Western Benin. Wageningen, Wageningen University. PhD thesis: 144.

Van Liere, M. J., E. A. Ategbo, J. Hoorweg, A. P. den Hartog and J. G. A. J. Hautvast (1994). The significance of socio-economic characteristics for adult seasonal body-weight fluctuations: a study in north-western Benin. *British Journal of Nutrition* 72: 479-488.

Van Liere, M. J., I. D. Brouwer and A. P. den Hartog (1996). A food ethnography of the Ottomari in north-western Benin: a systematic approach. *Ecology of Food and Nutrition* 34: 293-310.

Van Staveren, W. A. and M. C. Ocke (2001). Estimation of Dietary intake. *Present Knowledge in Nutrition* B. A. Bowman and R. M. Russel. Washington, ILSI press: 605-616.

Van Wijngaarden, J. P. (2005). The effect of in-school noodle consumption on out-school food consumption in children aged 6-8 year in rural North Vietnam. Wageningen, Wageningen University. Msc thesis: 67.

Vaughan, J. G., C. A. Geissler, B. E. Nicholson, E. Dowe and E. Rice (1997). *The new Oxford book of food plants.* Oxford, Oxford University Press.

Visser, M., L. C. De Groot, P. Deurenberg and W. A. Van Staveren (1995). Validation of dietary history method in a group of elderly women using measurements of total energy expenditure. *Br J Nutr* 74(6): 775-85.

Von Braun, J., H. Bouis, S. Kumar and R. Pandy-Lorch (1992). *Improving food security of the poor.* Washington DC, IFPRI.

Watier, B. (1982). *Un équilibre alimentaire en Afrique. Comment?* Neuilly sur Seine, Hoffman-La Roche.

Watson, J. L., Ed. (2006). *Golden arches east. Mc Donalds in East Asia.* Palo Alto, Stanford University Press.

Wheeler, E. F. and M. Abdullah (1988). Food allocations within the family: response to fluctuating food supply and food needs. *Coping with uncertainty in food supply and food needs.* I. De Garine and G. A. Harisson. Oxford, Clarendon Press: 437-451.

WHO (1995). Physical status: the use and interpretation of anthropometry. Report of a WHO Expert Committee. *World Health Organ Tech Rep Ser* 854: 1-452.

WHO (2003a). HIV and infant feeding: framework for priority action. from http://www.who.int/child-adolescent-health/New_Publications/NUTRITION/HIV_IF_Framework_pp.pdf.

WHO (2003b). Nutrient requirements for people living with HIV/AIDS. *Nutrient requirements for people living with HIV/AIDS. Report of a technical consultation.* , 13-15 May 2003, Geneva from http://www.who.int/nutrition/publications/Content_nutrient_requirements.pdf.

WHO (2004). Avian Influenza- Fact Sheet. Geneva, WHO.

WHO (2006). *WHO Child Growth Standards. Length/height-for-age, weight-for-age, weight-for-length, weight-for-height and body mass index-for-age. Methods and Development.* Geneva, World Health Organization from http://who.int/childgrowth/publications/technical_report_pub/en/index.html.

WHO and UNAIDS. (2005). Aids epidemic update. December 2005. from http://www.unaids.org/epi/2005/doc/EPIupdate2005_pdf_en/epi-update2005_en.pdf.

Willet, W. C. (1998). *Nutritional epidemiology.* New York, Oxford University Press.

Wishak, S.M. and S. van der Vynckt (1976). The use of nutritional "positive deviants" to identify approaches for modification of dietary practices. *Am J Publ Hlth* 66(1): 38-42,

Wolfe, W. S. and E. A. Frongillo (2001). Building household food security measurement tools from the ground up. *Food and Nutrition Bulletin* 22(1): 5-12.

Wollinka, O., E. Keely, R. Burkhalter and N. Bashir (1997). *Hearth Nutrition Model: applications in Haiti, Vietnam and Bangladesh.* Arlington, Basic Suppport for Institutionalizing Child Survival (BASICS) Project.

Womersley, J. and JGVA Durnin (1977). A comparison of the skinfold method with extent of 'overweight' and various weight-height relationships in the assessment of obesity. *Br J Nutr* 38(2): 271-84.

WorldBank (2000). *Cities in transition: a strategic view of urban and local government issues.* Washington DC, World Bank.

Yadav, S. K. and S. Sehgal (1995). Effect of home processing on total and extractable calcium and zinc content of spinach (Spinach oleracia) and amaranth (Amaranthus tricolor) leaves. *Plant Foods Hum Nutr* 48(1): 65-72.

Index

C

D

E

G

H

I

K

L

M

N

O

P

Q

R

S

Printed in the United States
by Baker & Taylor Publisher Services